Carpentry
Fourth Edition
WORKBOOK

AMERICAN TECHNICAL PUBLISHERS, INC.
HOMEWOOD, ILLINOIS 60430-4600

Thomas E. Proctor

Carpentry Workbook, 4th Edition, contains carpentry procedures commonly practiced in the trade. Specific procedures vary from job site to job site and must be performed by a qualified person. For maximum safety, always refer to specific manufacturer recommendations; job site requirements; applicable federal, state, and local regulations; and any authority having jurisdiction. The material contained is intended to be an educational resource for the user. American Technical Publishers, Inc. assumes no responsibility or liability in connection with this material or its use by any individual or organization.

4 5 6 7 8 9 – 04 – 9 8 7 6 5 4 3

Printed in the United States of America

ISBN 0-8269-0739-3

Contents

Unit Tests

Section Tests

Appendix

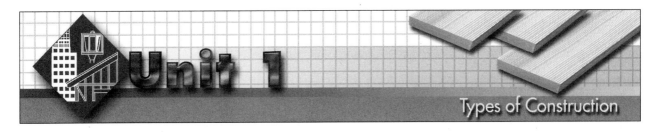

Name _____ Date _____

Multiple Choice

_____ **1.** Residential construction is a type of ___ construction.
       A. interior
       B. exterior
       C. light
       D. heavy

_____ **2.** The base of a mobile home rests on a ___ frame.
       A. steel
       B. plywood
       C. fractional
       D. exposed

_____ **3.** A ___ is not a prefabricated structural unit.
       A. box beam
       B. roof truss
       C. glued and laminated beam
       D. prehung door

_____ **4.** Three types of panel systems are ___, ___, and ___.
       A. open; closed; fold-out
       B. interior; exterior; bump-out
       C. upper; lower; bump-out
       D. light; heavy; look-out

_____ **5.** Post-and-beam construction is also known as ___ construction.
       A. pillar-and-beam
       B. plank-and-beam
       C. pillar-and-post
       D. none of the above

_____ **6.** A modular house is also known as a ___ house.
       A. fractional
       B. partial
       C. sectional
       D. proportional

_____ **7.** Specialization in the trade refers to carpenters working ___.
    A. in specific skill areas
    B. in specific locations
    C. on specific houses
    D. on specific days

_____ **8.** In post-and-beam construction, the beams are ___.
    A. never exposed
    B. generally exposed
    C. always exposed
    D. for appearance only

_____ **9.** Modular houses are ___% complete when delivered to the job site.
    A. 65
    B. 75
    C. 85
    D. 95

_____ **10.** Exterior walls of masonry buildings are built of ___.
    A. bricks
    B. blocks
    C. stone
    D. all of the above

True-False

T   F   **1.** The building of houses, condominiums, and small office buildings is light construction.

T   F   **2.** Alteration work is also known as remodeling.

T   F   **3.** Manufactured housing refers to homes that are built on site.

T   F   **4.** Residential construction employs more carpenters than any other construction area.

T   F   **5.** Carpenters build forms for pouring concrete.

T   F   **6.** Heavy construction uses reinforced concrete.

T   F   **7.** Mobile homes are available as single-, double-, and triple-wide.

T   F   **8.** Using precast units is more efficient than building forms and pouring concrete on the job site.

T   F   **9.** Over one-third of new residential construction today is manufactured housing.

T   F   **10.** Alteration work is a minor part of the carpentry trade in large cities.

T   F   **11.** Precast units are poured on the job site.

T   F   **12.** The inside surface of an exterior wall is left exposed in open panel systems.

T   F   **13.** Prefabricated units are manufactured in a shop and transported to the job site.

T   F   **14.** The wall sections are the basic units of a panel system.

T   F   **15.** Panel systems are normally constructed on the job site.

T   F   **16.** The printreading abbreviation for anchor bolt is AB.

Matching

_____ **1.** Monolithic

_____ **2.** Remodeling

_____ **3.** Residential construction

_____ **4.** Masonry construction

_____ **5.** Rebar

_____ **6.** Skyscrapers

_____ **7.** Brick veneer

_____ **8.** Heavy timber construction

_____ **9.** Trade specialization

_____ **10.** Mechanical core

A. outside walls of brick, stone, block, etc.

B. steel reinforcing bars

C. building bridges, trestles, piers, etc.

D. cast as a single piece

E. modular unit with kitchen and bathroom

F. type of light construction

G. alteration work

H. outside brick wall over wood stud wall

I. high-rise buildings

J. working in specific skill area

Short Answer

1. How does heavy construction differ from light construction?

_____

_____

_____

2. Discuss electrical and plumbing differences in open and closed panel systems.

_____

_____

_____

3. Discuss the advantages and disadvantages of manufactured housing.

_____

_____

_____

_____

Completion

_____ **1.** Carpentry is a trade of the ___ industry.

_____ **2.** ___ is the primary type of manufactured housing.

_____ **3.** Carpenters work on ___ construction and ___ construction.

_____ **4.** Brick-___ walls are built over framed wood stud walls.

_____ **5.** Wall units are cast on the floor and raised by crane in ___ construction.

_____ **6.** ___ construction includes the erection of buildings, bridges, railroad trestles, and similar structures.

_____ **7.** Subfloors for buildings of panel system construction are constructed on the ___.

_____ **8.** The most popular wood-framing method in light construction is ___ framing.

_____ **9.** Reinforced concrete contains ___ for strength.

_____ **10.** A framework is used in ___ high-rise buildings.

_____ **11.** Floor slabs are stack-cast around columns and raised by hydraulic jacks in ___ construction.

_____ **12.** Carpenters must complete the ___ work after modular units are set.

_____ **13.** In some parts of the United States, carpenters place ___ on walls and ceilings.

_____ **14.** Modular housing packages have a(n) ___ core with kitchen and bathroom equipment.

_____ **15.** Mobile homes are built in accordance with a building code established by the ___.

Name _____ Date _____

Short Answer

1. Briefly explain the type of work done by bricklayers.

_____

_____

_____

_____

2. Briefly explain the type of work done by cement masons.

_____

_____

_____

_____

3. Briefly explain the type of work done by electricians.

_____

_____

_____

_____

4. Briefly explain the type of work done by millwrights.

_____

_____

_____

5. Briefly explain the type of work done by operating engineers.

_____

_____

_____

_____

6. Briefly explain the type of work done by plumbers.

_____

_____

_____

_____

_____

7. Briefly explain the type of work done by drywallers.

_____

_____

_____

_____

8. Briefly explain the type of work done by construction craft laborers.

_____

_____

_____

_____

9. Briefly explain the type of work done by structural-steel workers.

_____

_____

_____

_____

10. Briefly explain the type of work done by pipefitters.

_____

_____

_____

_____

_____

11. Briefly explain the type of work done by plasterers.

_____

_____

_____

_____

12. Briefly explain the type of work done by painters and paperhangers.

_____

_____

_____

_____

13. Briefly explain the type of work done by roofers.

_____

_____

_____

_____

14. Briefly explain the type of work done by sheet-metal workers.

_____

_____

_____

_____

15. Briefly explain the type of work done by glaziers.

_____

_____

_____

_____

_____

16. What is the function of the Associated General Contractors (AGC)?

_____

_____

_____

_____

_____

17. What groups does the National Association of Home Builders (NAHB) represent?

_____

_____

_____

_____

_____

18. What points do the United Brotherhood of Carpenters and Joiners of America (UBC) negotiate with contractors' associations?

_____

_____

_____

_____

_____

_____

_____

19. How can a young person prepare for the carpentry trade?

_____

_____

_____

_____

20. Why is carpentry considered a key element on most construction jobs?

_____

_____

_____

_____

Completion

_____ **1.** The printreading abbreviation for bathtub is ___.

_____ **2.** A trade association is an organization representing the ___ of specific products.

_____ **3.** The National Hardwood Lumber Association (NHLA) publishes technical information on lumber grading, ___, and harvesting.

_____ **4.** The ___ promotes the use of wood by ensuring that wood products are widely accepted by model codes and regulations.

_____ **5.** The ___ has the goal of improving and expanding the uses of Portland cement and concrete.

_____ **6.** APA—The Engineered Wood Association is a(n) ___ trade association of the United States and Canadian engineered wood products industry.

_____ **7.** The California Redwood Association (CRA) is the ___ for redwood lumber producers.

_____ **8.** The ___ is the national technical trade association of the structural glulam timber industry.

_____ **9.** The Occupational Safety and Health Administration (OSHA) is a federal agency that requires all ___ to provide a safe environment for their employees.

_____ **10.** Southern Forest Products Association members produce about ___% of the southern pine lumber in the United States.

_____ **11.** The ___ is the carpenters' union.

Math—Adding Whole Numbers

1.
```
      7
     13
    901
     72
+  1081
```

2.
```
     8396
+  14,511
```

3.
```
   638
   711
   461
+  358
```

4.
```
  45,681
  56,238
+     57
```

5.
```
     855
    4695
+  15,874
```

Math—Subtracting Whole Numbers

1.
```
   98
-  61
```

2.
```
   936
-  231
```

3.
```
   7003
-  4938
```

4.
```
   63,889
-  50,004
```

5.
```
  124,578
-   6584
```

Math—Multiplying Whole Numbers

1.
$$\begin{array}{r} 45 \\ \times\ 12 \\ \hline \end{array}$$

2.
$$\begin{array}{r} 265 \\ \times\ 381 \\ \hline \end{array}$$

3.
$$\begin{array}{r} 3005 \\ \times\ 101 \\ \hline \end{array}$$

4.
$$\begin{array}{r} 420 \\ \times\ 89 \\ \hline \end{array}$$

5.
$$\begin{array}{r} 6284 \\ \times\ 154 \\ \hline \end{array}$$

Math—Dividing Whole Numbers

1. $288 \div 12 =$

2. $456 \div 4 =$

3. $141{,}912 \div 438 =$

4. $3456 \div 8 =$

5. $2025 \div 45 =$

Printreading Symbols

Identify the building
material shown.

_____

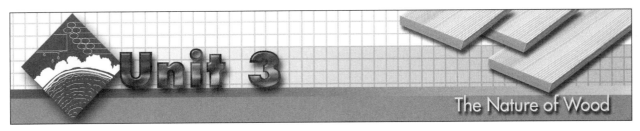

Name _____ Date _____

Completion

_____ **1.** A tree grows by forming new wood ___.

_____ **2.** The light-colored section of an annual ring is ___.

_____ **3.** Water is present in ___ cavities and ___ walls of wood.

_____ **4.** The moisture content of wood used for interior finish materials should be between ___% and ___%.

_____ **5.** Wood ___ in strength as moisture content ___.

_____ **6.** ___ is the growing portion of a tree.

_____ **7.** Lumber should have a moisture content ___ with the air that will surround it after it is installed.

_____ **8.** Wood cell walls are composed of ___ matter.

_____ **9.** When water leaves the cell walls, the cells begin to ___ in size, causing the wood to shrink.

_____ **10.** Wood shrinks more ___ the grain than ___ the grain.

Multiple Choice

_____ **1.** Wood does not decay when its moisture content is below ___%.
   A. 20
   B. 24
   C. 28
   D. 32

_____ **2.** Wood cells attain a length of approximately ___.
   A. ⅛″ for softwood and hardwood
   B. ⅛″ for softwood and ¼₄″ for hardwood
   C. ¼₄″ for softwood and hardwood
   D. ¼₄″ for softwood and ⅛″ for hardwood

_____ **3.** As lumber dries out, the water evaporates from the cell ___.
  A. walls first and cell cavities next
  B. cavities first and cell walls next
  C. walls and cell cavities at the same time
  D. none of the above

_____ **4.** When lumber reaches the equilibrium moisture content, ___.
  A. growth stops
  B. shrinkage stops
  C. growth begins
  D. shrinkage begins

_____ **5.** The light-colored wood directly beneath the cambium is ___.
  A. springwood
  B. summerwood
  C. heartwood
  D. sapwood

_____ **6.** The glue bond of plywood improves as the moisture ___.
  A. increases
  B. decreases
  C. remains stable
  D. none of the above

_____ **7.** New cell growth of a tree is formed in the ___.
  A. lignin
  B. cellulose
  C. cambium
  D. none of the above

_____ **8.** Medullary rays ___ food.
  A. store and transport
  B. consume and transport
  C. consume and divert
  D. store and convert

_____ **9.** The fiber saturation point is reached when the cell ___.
  A. walls contain water and cavities contain water
  B. walls contain no water and cavities contain water
  C. walls contain no water and cavities contain no water
  D. walls contain water and cavities contain no water

_____ **10.** During tree growth, the number of annual rings ___.
  A. does not change
  B. decreases one per year
  C. increases one per year
  D. increases two per year

Matching

_____ **1.** Springwood

_____ **2.** Bark

_____ **3.** Equilibrium moisture content

_____ **4.** Sapwood

_____ **5.** Summerwood

_____ **6.** Sap

_____ **7.** Rot

_____ **8.** Pith

_____ **9.** Cells

_____ **10.** Heartwood

A. watery fluid

B. light wood of a tree

C. decay

D. light-colored section of annual ring

E. outer tree covering

F. dark-colored section of annual ring

G. central core of a tree

H. dark wood of a tree

I. lumber stops shrinking

J. fibers

Identification—Tree Growth

_____ **1.** Bark

_____ **2.** Sapwood

_____ **3.** Heartwood

_____ **4.** Pith

_____ **5.** Cambium

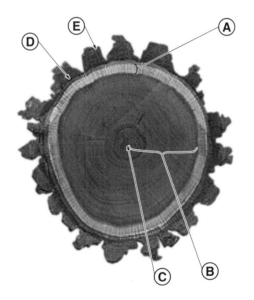

True-False

T   F   **1.** Wood is composed of numerous cells.

T   F   **2.** Annual rings are narrower during dry seasons.

T   F   **3.** Springwood is normally weaker than summerwood.

T   F   **4.** Lignin holds cells together.

T F   **5.** Trees have one bark layer.

T F   **6.** An annual ring represents one decade of cellular growth.

T F   **7.** Heartwood lumber is normally stronger than sapwood lumber.

T F   **8.** A tree stops growing when no new cells are formed.

T F   **9.** Heartwood is darker than sapwood.

T F   **10.** The dark-colored section of an annual ring is springwood.

T F   **11.** The printreading abbreviation for beam is BE.

Short Answer

1. Describe the steps involved in determining the moisture content of a piece of wood by the oven-drying method.

_____

_____

_____

_____

_____

_____

2. How does a moisture meter give a reading of the moisture content of wood?

_____

_____

3. What causes the sapwood in the central part of a tree to turn into heartwood?

_____

_____

_____

4. Discuss how wood shrinks in relation to grain direction.

_____

_____

_____

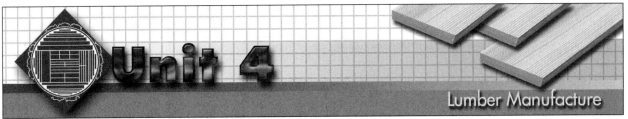

Unit 4

Lumber Manufacture

Name _____ Date _____

Matching

_____ **1.** Medium knot

_____ **2.** Check

_____ **3.** Kiln

_____ **4.** Dried

_____ **5.** Dry rot

_____ **6.** Creosote

_____ **7.** Knot

_____ **8.** Split

_____ **9.** Combustible

_____ **10.** Shake

A. separation of wood fibers across annual growth rings

B. most common fungus damage

C. type of wood preservative

D. ignites and burns easily

E. check extending completely through the lumber

F. separation of wood fibers between annual growth rings

G. over ¾″ and less than 1½″ in diameter

H. most common natural wood defect

I. oven

J. seasoned

Multiple Choice

_____ **1.** Approximately ___ of a log becomes usable construction lumber.
  A. one-fourth
  B. one-third
  C. one-half
  D. two-thirds

_____ **2.** Factors that determine the grade of lumber include ___.
  A. type and place grown
  B. length and width of boards
  C. number and type of defects
  D. all of the above

_____ **3.** Dry rot in wood is caused by a ___.
  A. lack of water during the growing season
  B. low humidity level
  C. fungus
  D. all of the above

_____ **4.** ___ is a method for cutting logs into lumber.
  A. Rift sawing
  B. Plainsawing
  C. Quartersawing
  D. all of the above

_____ **5.** A ___ is a common warpage shape of lumber.
  A. split
  B. knot
  C. bow
  D. none of the above

Identification—Lumber

_____ **1.** Crown

_____ **2.** Cup

_____ **3.** Bow

_____ **4.** Quartersawn

_____ **5.** Plainsawn

_____ **6.** Twist

(A)

(B)

(C)

(D)

(E)

(F)

Completion

_____ **1.** The portion of a log that does not become lumber is known as ___.

_____ **2.** Hardwood lumber is generally ___ dried.

_____ **3.** Softwood lumber is generally ___ dried.

_____ **4.** A large knot is ___″ or more in diameter.

_____ **5.** A pin knot is ___″ or less in diameter.

_____ **6.** Dry rot can only survive in wood with a moisture content of ___% or more.

_____ **7.** Most lumber is produced by ___ sawing.

_____ **8.** Two methods of drying lumber are ___ drying and ___ drying.

_____ **9.** ___ is a board's deviation from a flat plane, edge to edge.

_____ **10.** ___ that may mar lumber appearance include chipped grain, surface skip mars, and knife marks.

_____ **11.** The printreading abbreviation BR stands for ___.

Short Answer

1. How is lumber air dried?

_____

_____

_____

_____

2. What are the advantages of quartersawn lumber?

_____

_____

_____

3. Discuss the advantages of plainsawn lumber.

_____

_____

_____

4. How is residue from sawmilling operations utilized today?

_____

_____

_____

5. Discuss efforts being made to conserve trees.

_____

_____

_____

6. How is lumber kiln dried?

_____

_____

_____

_____

Softwood and Hardwood

Name _____ Date _____

True-False

T   F   **1.** Douglas fir is mostly used for rough construction in the western part of the United States.

T   F   **2.** Structural light framing lumber applications include load-bearing members.

T   F   **3.** Studs are 2″ × 4″ or 2″ × 6″ pieces of lumber 12′ or shorter.

T   F   **4.** Softwood is normally more expensive than hardwood.

T   F   **5.** Yellow poplar is a softwood.

T   F   **6.** Southern pine is the predominant rough construction wood used in the southeastern part of the United States.

T   F   **7.** Approximately 25% of total lumber production is hardwood.

T   F   **8.** FAS indicates that hardwood is first and second grade.

T   F   **9.** Finish materials are used for interior trim work only.

T   F   **10.** The printreading abbreviation BK stands for benchmark.

Completion

_____ **1.** ___ and ___ are the two main classes of trees.

_____ **2.** ___ lumber is used for rough construction.

_____ **3.** The grade of lumber is based on ___, ___, and ___.

_____ **4.** ___ classification of lumber is used for beams, stringers, posts, and timbers.

_____ **5.** Three main softwood end-use categories are ___, ___, and ___.

_____ **6.** Most hardwood trees grow in the ___ part of the United States.

_____ **7.** Evergreen trees having needles and cones are ___.

_____ **8.** Timbers are no smaller than ___″ wide by ___″ thick.

_____ **9.** C select lumber contains small, tight ___.

_____ **10.** ___ trees have broad leaves that fall off in the autumn.

Identification—Grade Marks

_____ **1.** Mill identification number

_____ **2.** Lumber grade

_____ **3.** Wood species

_____ **4.** Condition of seasoning when surfaced

_____ **5.** Association certification trademark

12 **STAND** S-DRY

Ⓔ Ⓐ Ⓓ Ⓒ Ⓑ

Western Wood Products Association

Multiple Choice

_____ **1.** ___ marks are stamped on lumber to provide grading information.
A. Inspection
B. Quality
C. Grade
D. none of the above

_____ **2.** ___ is not a hardwood.
A. Birch
B. Gum
C. Redwood
D. Basswood

_____ **3.** Over ___% of the wood used for construction is softwood.
A. 25
B. 50
C. 75
D. 90

Printreading Symbols

Identify this symbol.

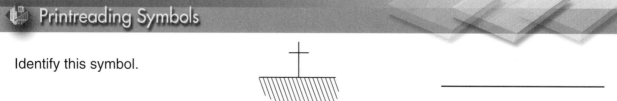

_____

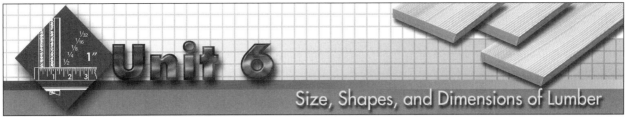

Unit 6
Size, Shapes, and Dimensions of Lumber

Name _____ Date _____

Abbreviations

_____ **1.** Surfaced on one side

_____ **2.** Surfaced on two sides

_____ **3.** Surfaced on four sides

_____ **4.** Surfaced on two edges

_____ **5.** Surfaced on one edge

_____ **6.** Saw sized

_____ **7.** Surfaced on two sides and one edge

_____ **8.** Surfaced on one side and two edges

_____ **9.** Surfaced on one side and one edge

Math

_____ **1.** A board measuring 1″ × 4″ × 6′ contains ___ BF.

_____ **2.** Two boards measure 2″ × 6″ × 6′. They measure ___ BF.

_____ **3.** Cherry costs $2.25 per BF for 1″ nominal thickness (¹³⁄₁₆″ actual thickness). How much will three pieces 6″ wide and 8′ long cost?

_____ **4.** A lumber order calls for the following:
   120 lineal feet of ¾″ screen mold @ $.28 per lineal foot
   60 lineal feet of 1″ × 4″ pine @ $.67 per lineal foot
   3 sheets of ¼″ luan @ $18.16 per sheet
What is the total material cost for this lumber order?

_____ **5.** An order for hardwood lumber calls for the following:
   150 BF of ¹³⁄₁₆″ Appalachian cherry @ $2.25 per BF
   75 BF of ½″ plainsawn oak @ $1.75 per BF
   The sales tax on this order is 6%. Also, there is a 10% delivery charge. (*Note:* Do not figure sales tax on the delivery charge.)
What is the total cost of this lumber?

_____ **6.** A contractor places an order on June 16th for two hundred 2″ × 4″ × 8′ studs at $1.62 each, and thirty 1″ × 6″ × 16′ pine boards at $680.00 per thousand board feet. The lumberyard offers free delivery for orders exceeding $500.00 and charges 5% delivery charge for orders less than $500.00. The lumberyard also allows a 2% discount if payment is made within 10 days of invoice date. The contractor pays the order on June 24th. What is the total amount paid? (_Note:_ There is no sales tax on this order.)

_____ **7.** In problem #6, would the contractor have been better off by increasing the order by three 1″ × 6″ × 16′ pine boards?

_____ **8.** A contractor is adding a deck onto an existing house. Enough material remains from a previous job to complete this job except for forty-two 2″ × 4″ × 16′ pieces of treated yellow pine. What is the additional material cost at $.28 per lineal foot?

_____ **9.** A remodeling contractor has a job that calls for a 212 sq ft entry hall floor to be covered with 6″ × 6″ oak parquet flooring, which costs $1.39 per piece. Carpentry charges for laying this floor are $16.48 per hour and four hours are required. What is the total cost the contractor must pay for materials and labor on this job?

_____ **10.** In problem #9, the contractor adds 30% to the labor and material costs in order to meet overhead and make a fair profit. What is the total invoice amount to the customer?

_____ **11.** A piece of walnut is 3″ thick, 16″ wide, and 4′ long. How many board feet does it contain?

_____ **12.** Three pieces of 1″ lumber 8″ wide and 8′ long contain ___ BF.

_____ **13.** A stack of lumber contains the following:
　　three 1″ × 6″ × 8′ pieces of yellow pine
　　six 2″ × 10″ × 12′ pieces of redwood
　　six 1″ × 8″ × 10′ pieces of fir
How many board feet of lumber are in this stack?

_____ **14.** Cherry costs $3.46 per BF. How much will 132 BF of cherry cost?

_____ **15.** Sixteen board feet of maple at $4.02 per BF and three sheets of ¼″ luan mahogany at $18.32 per sheet will cost how much?

_____ **16.** How many board feet are needed to glue a walnut top that is ¾″ thick, 24″ wide, and 54″ long?

_____ **17.** A project requires 32 BF of 1″ cherry and 18 BF of 1″ oak. What percentage of the total wood required is oak?

_____ **18.** At $1.47 per BF, what is the cost of 172 BF of lumber?

_____**19.** A piece of lumber contains 4 BF. Which of the following could not be this piece?
    A. 1″ × 12″ × 48″
    B. 2″ × 6″ × 48″
    C. 4″ × 6″ × 24″
    D. 9″ × 2″ × 48″

_____**20.** A 1″ × 1″ × 8′ strip of lumber has ___ BF.

Multiple Choice

_____ **1.** Softwood lumber is usually sold in even lengths ranging from ___′ to ___′.
    A. 4; 16
    B. 6; 24
    C. 4; 26
    D. 6; 16

_____ **2.** The actual size of a 2″ × 4″ piece of lumber is ___″ × ___″.
    A. 1⅝; 3⅝
    B. 1½; 3½
    C. 1⅝; 3½
    D. 1½; 3⅝

_____ **3.** Lumber measurements are stated in the following order: ___.
    A. thickness, width, length
    B. thickness, length, width
    C. width, thickness, length
    D. length, thickness, width

_____ **4.** Metric lumber and panel sizes are based on the actual standard lumber sizes and are expressed in ___.
    A. meters
    B. centimeters
    C. millimeters
    D. U.S. Customary

_____ **5.** A piece of lumber that measures 2″ × 8″ × 16′ contains ___ BF.
    A. 14½
    B. 16¼
    C. 21⅓
    D. 24⅔

_____ **6.** Resawn lumber is generally used for ___.
    A. exterior trim
    B. siding
    C. paneling
    D. all of the above

Identification—Lumber Dimensions

_____ **1.** Width

_____ **2.** Thickness

_____ **3.** Length

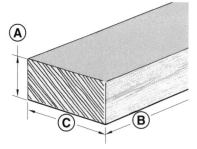

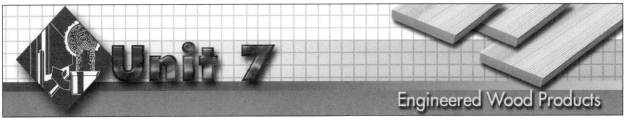

Engineered Wood Products

Name _____ Date _____

Matching

_____ **1.** Glulam

_____ **2.** Wood I-joists

_____ **3.** Trademark

_____ **4.** Cross-laminated

_____ **5.** Mill number

_____ **6.** Plyform®

_____ **7.** Ply

_____ **8.** Core

_____ **9.** Face and back veneer

_____ **10.** Peeler log

A. descriptive nomenclature of panel

B. location where panel was manufactured

C. wood laminations bonded with adhesive

D. single veneer sheet

E. center layer

F. web between top and bottom flanges

G. panel for concrete forms

H. outside layers of a panel

I. used to make veneer

J. layers at right angles

True-False

T   F   **1.** Plyform® has surface protection resulting in a durable forming surface.

T   F   **2.** Each layer of plywood consists of one or more plies.

T   F   **3.** Fiberglass-reinforced plastic plywood has glass fiber mats used as the center plies in plywood.

T   F   **4.** Interior panels may be used for exterior sheathing purposes.

T   F   **5.** Oriented strand board panels may be used for sheathing.

T   F   **6.** Structural plywood panels are manufactured from hardwood.

T   F   **7.** Plywood always has an odd number of layers.

T   F   **8.** Composite panels offer the advantage of structural panels with an appearance-grade finish.

T   F   **9.** Medium density fiberboard is a structural panel.

T	F	**10.** Particleboard is commonly used for floor underlayment.
T	F	**11.** The thickness measurement given in the trademark refers to nominal panel thickness.
T	F	**12.** The plies of interior panels are bonded together with moisture-resistant or waterproof glue.
T	F	**13.** Nonstructural wood panels are used for sheathing only.
T	F	**14.** Hardboard panels are generally 42″ wide.
T	F	**15.** Exposure 1 panels are recommended where panels will have short periods of exposure to moisture during construction.
T	F	**16.** The printreading abbreviation for board is BD.

Identification—APA Trademark

_____ **1.** Panel grade

_____ **2.** Tongue-and-groove

_____ **3.** Thickness

_____ **4.** Exposure durability classi-fication

_____ **5.** Performance rated panel standard of the APA— The Engineered Wood Association

_____ **6.** Mill number

APA
THE ENGINEERED WOOD ASSOCIATION

(F) RATED STURD-I-FLOOR (A)
24 OC 23/32 INCH
SIZED FOR SPACING (B)
(E) T&G NET WIDTH 47-1/2
EXPOSURE 1 (C)
000
PS 1-95 UNDERLAYMENT
PRP-108 (D)

Completion

_____ **1.** A widely used structural wood panel is ___.

_____ **2.** Adjacent layers of a plywood panel are placed at a(n) ___° angle to one another.

_____ **3.** During manufacture of plywood, ___ are used to remove small defects.

_____ **4.** Medium density overlay (MDO) and high density overlay (HDO) plywood panels have ___ resin-treated fiber overlays.

_____ **5.** Thin layers of wood that form plywood are known as ___.

_____ **6.** Oriented strand board panels normally have ___ to ___ layers.

_____ **7.** Particleboard is available in thicknesses from ___″ to ___″.

_____ **8.** HDO plywood concrete forms can be reused ___ to ___ times.

_____ **9.** Oriented strand board (OSB) panels are ___ expensive than plywood panels.

_____ **10.** APA rated sheathing is used where the highest ___ and ___ are required.

_____ **11.** The APA recommends ___ panels for continuous weather or moisture exposure.

_____ **12.** Four grades of hardboard are ___, ___, ___, and ___.

_____ **13.** ___ is a combination subfloor/underlayment panel used under a carpet and pad.

_____ **14.** ___ plywood panels are primarily used for interior finish and cabinetwork.

_____ **15.** ___ is manufactured using solid wood members or veneers, wood strands and fibers, or a combination of solid wood members and wood strand members.

Multiple Choice

_____ **1.** Laminated veneer lumber is used for ___.
    A. beams, headers, and rafters
    B. underlayment and subfloors
    C. concrete formwork
    D. all of the above

_____ **2.** The outside layers of a panel are known as the ___ and ___.
    A. front veneer; back veneer
    B. front veneer; rear veneer
    C. face veneer; rear veneer
    D. face veneer; back veneer

_____ **3.** Veneers are graded according to their ___.
    A. appearance
    B. size and number of repairs made during manufacture
    C. natural growth characteristics
    D. all of the above

_____ **4.** Plywood was first manufactured in the United States near the ___.
    A. middle of the nineteenth century
    B. end of the nineteenth century
    C. beginning of the twentieth century
    D. middle of the twentieth century

_____ **5.** OSB rim boards fill the space between a ___ and the bottom plate of a wall.
    A. rafter
    B. joist
    C. sill plate
    D. all of the above

_____ **6.** Plyform® panel grades include ___ and ___.
     A. appearance; A-smooth
     B. Class I; Class II
     C. interior; exterior
     D. none of the above

_____ **7.** Parallel and oriented strand lumber consist of long wood strands ___ along the length of the lumber.
     A. used as plies
     B. oriented
     C. and sawdust
     D. all of the above

_____ **8.** Standard sizes of plywood panels are ___, ___, and ___.
     A. 2′ × 6′; 3′ × 6′; 4′ × 8′
     B. 3′ × 8′; 4′ × 8′; 4′ × 10′
     C. 4′ × 8′; 4′ × 10′; 5′ × 12′
     D. 4′ × 8′; 4′ × 10′; 4′ × 20′

_____ **9.** A plywood panel with a grade rating of 32/16 can be used for ___.
     A. sheathing over 32″ OC roof rafters and for subflooring over 16″ OC joists
     B. sheathing over 16″ OC roof rafters and for subflooring over 32″ OC joists
     C. sheathing over 32″ OC roof rafters with 16″ nailing
     D. subflooring over 16″ OC joists when not over 32″ from any one edge

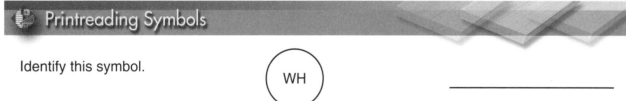

Printreading Symbols

Identify this symbol.

(WH)

_____

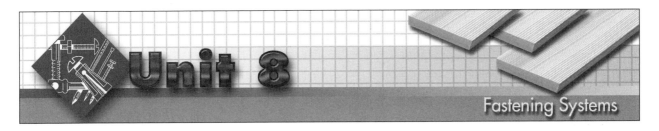

Name _____ Date _____

Multiple Choice

_____ **1.** Mastics have a(n) ___ base.
  A. asphalt
  B. rubber
  C. resin
  D. all of the above

_____ **2.** Drive pins and studs are normally placed into concrete with ___ tools.
  A. hand
  B. air-operated
  C. electric
  D. powder-actuated

_____ **3.** Framing members will normally pull loose at their joints before ___.
  A. rupturing
  B. breaking in their length
  C. breaking in their cross section
  D. none of the above

_____ **4.** Duplex nails are used primarily for ___ construction.
  A. heavy
  B. light
  C. temporary
  D. permanent

_____ **5.** Duplex nails are also known as ___ nails.
  A. double-headed
  B. finish
  C. casing
  D. none of the above

_____ **6.** ___ screws are used to expand light-duty lead-alloy anchors.
  A. Lag
  B. Wood
  C. Sheet-metal
  D. all of the above

_____ **7.** Stove bolts normally have ___ heads.
      A. slotted flat or round
      B. square
      C. triangular
      D. oval

_____ **8.** A(n) ___d nail is 2½″ in length.
      A. 4
      B. 6
      C. 8
      D. 12

Completion

_____ **1.** The ___ nail, while similar to a common nail, has a smaller head and thinner shank.

_____ **2.** Wire brads are available from ___″ to ___″ in length.

_____ **3.** ___ nails are used most often for wood-frame construction.

_____ **4.** ___ nails are made of hardened steel.

_____ **5.** A ___ is used to drive finish nails below the surface.

_____ **6.** Recessed cross slot screw heads are commonly called ___ head screws.

_____ **7.** Heads on wood screws are flat, round, or ___.

_____ **8.** Wood screws are available from ___ to ___ in length.

_____ **9.** ___ screws cut threads in metal when driven.

_____ **10.** Wood screw heads have a single slot, a square recess, or a ___ slot.

_____ **11.** Shank and ___ holes are drilled to receive lag bolts.

_____ **12.** Stove bolts are commercially available from ___″ to ___″ in length.

_____ **13.** When a nut and bolt are used in wood, a ___ should also be used.

_____ **14.** ___ bolts or ___ bolts are used to expand the sides of expansion shields.

True-False

T   F   **1.** Nail sizes are designated by a number and the letter _p_.

T   F   **2.** Finish nails are thicker than common nails.

T   F   **3.** Nails derive their holding power from the pressure of wood fibers and different types of shanks.

T	F	**4.** Nails have greater holding power than screws.
T	F	**5.** Finish nails are most often used to fasten exterior insulation boards.
T	F	**6.** Wood screw diameters are identified by gauge numbers.
T	F	**7.** Box nails bend more easily than common nails.
T	F	**8.** Stove bolts have a square shoulder below the head.
T	F	**9.** A screw anchor is also known as a toggle bolt.
T	F	**10.** Carriage bolts are used only for fastening steel parts together.
T	F	**11.** A toggle bolt is also known as an expansion anchor.
T	F	**12.** Machine bolts are used to fasten wood or metal parts together.
T	F	**13.** A hole slightly smaller than the plug diameter must be drilled for expansion shields.
T	F	**14.** Lag bolts have machine threads and square or hexagonal heads.
T	F	**15.** Toggle bolts have machine screw threads.
T	F	**16.** The printreading abbreviation for building line is BU.

Identification—Nails and Bolts

_____ 1. Flat-head stove bolt

_____ 2. Machine bolt

_____ 3. Carriage bolt

_____ 4. Lag bolt

_____ 5. Round-head stove bolt

_____ 6. Round shank nail

_____ 7. Barbed nail

_____ 8. Longitudinally grooved nail

_____ 9. Screw nail

_____ 10. Spiral nail

_____ 11. Oval nail

_____ 12. Triangle nail

_____ 13. Square nail

_____ 14. Annular nail

Matching

_____ **1.** Deformed shanks

_____ **2.** Screw

_____ **3.** Urea resin

_____ **4.** Adhesives

_____ **5.** Staples

_____ **6.** Bolts

_____ **7.** Contact cement

_____ **8.** Stud-bolt anchors

_____ **9.** Nail

_____ **10.** d

A. used for lighter applications

B. adhesive for plastic laminates

C. expansion plug, sleeve, and wedge

D. barbs, spirals, and rings

E. threaded fastener

F. powdered plastic resin glue

G. threadless fastener

H. glues and mastics

I. used where removal of fastener is needed

J. nail size designation

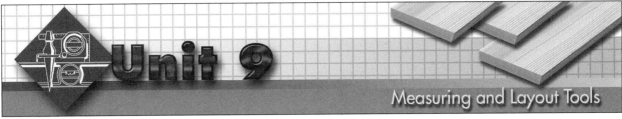

Name _____ Date _____

Completion

_____ **1.** The blade of a framing square is ___″ wide and ___″ long.

_____ **2.** A Speed® Square is used to mark a ___° or ___° angle.

_____ **3.** A laser hand level can be mounted on a ___ or placed on a flat surface.

_____ **4.** The term *plumb* refers to a ___ position.

_____ **5.** The ___ vial of a carpenter's level is used for leveling.

_____ **6.** The most commonly used angle in the building trades is ___°.

_____ **7.** Combination squares usually have adjustable blades ___″ long.

_____ **8.** The Essex board measure table is located on the back of the ___ of a framing square.

_____ **9.** A(n) ___ is used to lay out joints that meet at angles other than 90°.

_____ **10.** Tape measures may be graduated in English and/or ___ measurements.

_____ **11.** The printreading abbreviation for cabinet is ___.

Identification—Combination Square

_____ **1.** Thumbscrew

_____ **2.** Scriber

_____ **3.** Head

_____ **4.** Blade

_____ **5.** Leveling vial

Matching

_____ **1.** Level

_____ **2.** Combination square

_____ **3.** Angle divider

_____ **4.** Plumb

_____ **5.** Chalk line reel

_____ **6.** Sliding T-bevel

_____ **7.** Square gauges

_____ **8.** Trammel points

_____ **9.** Tape measure

_____ **10.** Scriber

A. pertaining to a vertical plane

B. used to snap lines on flat surfaces

C. used to lay out circles of any size

D. used for transferring angles

E. used for marking angles other than 90°

F. pertaining to a horizontal plane

G. used for marking 45° and 90° angles

H. used with framing square to lay out roof rafters and stair stringers

I. has two legs

J. used more often than any other measuring tool

Math

1.
$$\begin{array}{r} 138 \\ \times\ 21 \\ \hline \end{array}$$

4.
$$\begin{array}{r} 13'\text{-}6'' \\ -\ 10'\text{-}8'' \\ \hline \end{array}$$

2.
$$\begin{array}{r} 1'\text{-}2'' \\ 3'\text{-}6'' \\ +\ 10'\text{-}3'' \\ \hline \end{array}$$

5.
$$\begin{array}{r} 36'\text{-}6'' \\ -\ 18'\text{-}9'' \\ \hline \end{array}$$

6. $32\overline{)536}$

3. $26\overline{)5226}$

7. $^{10}/_{16} = /_8$

8. $^3/_8 + ^1/_2 + ^5/_{16} =$

9. $^3/_4 \times ^1/_2 =$

10. $^{12}/_{16} = /_4$

11. $^5/_8 + 2 =$

12. $^3/_4$ of $3' =$

13. $^3/_5$ of $15 =$

14. $75'' = \quad '- \quad ''$

15. $16' \div 16'' =$

16.
$$\begin{array}{r} 7'\text{-}5'' \\ 21'\text{-}6'' \\ +\quad 13'\text{-}8'' \\ \hline \end{array}$$

17. $12^3/_4 + 5^1/_8 + 16^1/_2 =$

18.
$$\begin{array}{r} 18'\text{-}\ 9'' \\ -\quad 12'\text{-}10'' \\ \hline \end{array}$$

19. $^3/_8 \times 12 =$

20.
$$\begin{array}{r} 3'\text{-}\ 6'' \\ 13'\text{-}\ 4'' \\ +\quad 2'\text{-}11'' \\ \hline \end{array}$$

Name _____ Date _____

Completion

_____ 1. The size of a hammer is determined by the ___ of its ___.

_____ 2. The ___-claw hammer is used primarily for finish work.

_____ 3. The ___-claw hammer is used primarily for rough work.

_____ 4. A(n) ___ is used to sink nail heads below the surface.

_____ 5. Nails have more holding power when driven ___ the grain.

_____ 6. The face of a hammerhead may be ___ shaped or ___ shaped.

_____ 7. Hammer handle materials include ___, ___, and ___.

_____ 8. Nails should be driven at a(n) ___ to increase the holding power between two pieces fastened face-to-face.

_____ 9. Nails should be ___ near the end of a board to prevent splitting.

_____ 10. A(n) ___-faced hammer can drive nails flush without marring the wood surface of the lumber.

_____ 11. Two types of surfaces on hammerhead faces are ___ and ___.

_____ 12. Nail set sizes range from ___″ to ___″.

_____ 13. The ___ hatchet has a curved face to dimple the surface.

_____ 14. Staplers used on the job are generally provided by the ___.

_____ 15. ___ injuries are the most frequently occurring injuries when using hammers and hatchets.

_____ 16. A(n) ___ hatchet is used for installing roof shingles.

_____ 17. Screwdrivers with long blades allow greater ___ to be applied.

_____ 18. Three types of staplers are the ___ tacker, ___ tacker, and ___ tacker.

_____ 19. A(n) ___-hatchet is used when building wood forms for concrete.

_____ 20. When recessing flat-head screws, the shank hole must be ___.

_____ **21.** When using an adjustable wrench, pressure should be applied against the ___ jaw.

_____ **22.** A wallboard hatchet is also known as a(n) ___.

_____ **23.** The size of a screwdriver is determined by the ___ of its ___.

_____ **24.** Three basic types of screwdrivers are the ___, ___, and ___.

_____ **25.** A(n) ___ chisel has a nail slot at one end for pulling nails in tight areas.

_____ **26.** A ripping bar is also known as a(n) ___ bar.

_____ **27.** Wrenches tend to slip more often when a(n) ___ pressure is applied.

_____ **28.** Tips of standard screwdrivers range from ___″ to ___″ in width.

_____ **29.** Ripping bars are commercially available in lengths from ___″ to ___″.

_____ **30.** The printreading symbol for ___ is CSMT.

Identification—Hammer

_____ **1.** Cheek

_____ **2.** Adze eye

_____ **3.** Neck

_____ **4.** Handle

_____ **5.** Claw

_____ **6.** Face

_____ **7.** Poll

_____ **8.** Head

Math—Adding Fractions

1. $1\frac{7}{8}'' + 2\frac{3}{8}'' + 4\frac{5}{8}'' =$

2. $3\frac{3}{4}'' + 1'\text{-}2\frac{1}{4}'' =$

3. $5\frac{1}{8}'' + 1\frac{7}{8}'' + 3'\text{-}6\frac{3}{8}'' =$

4. $3\frac{13}{16}'' + \frac{7}{8}'' =$

5. $1\frac{1}{4}'' + \frac{5}{16}'' + \frac{1}{4}'' =$

6. $3\frac{1}{2}'' + 1\frac{1}{2}'' + \frac{11}{16}'' =$

7. $\frac{7}{16}'' + \frac{7}{8}'' + 1\frac{3}{16}'' =$

8. $1\frac{1}{4}'' + 1\frac{1}{2}'' + \frac{3}{4}'' =$

9. $\frac{7}{8}'' + \frac{5}{16}'' + \frac{3}{4}'' + \frac{1}{2}'' =$

10. $\frac{9}{16}'' + \frac{5}{8}'' + \frac{3}{4}'' + \frac{1}{2}'' =$

Math—Subtracting Fractions

1. $\frac{3}{4}'' - \frac{1}{4}'' =$

2. $\frac{13}{16}'' - \frac{5}{16}'' =$

3. $\frac{7}{8}'' - \frac{1}{2}'' =$

4. $2\frac{3}{4}'' - \frac{3}{8}'' =$

5. $5\frac{5}{8}'' - 2\frac{1}{2}'' =$

6. $1'\text{-}3\frac{1}{8}'' - 1'\text{-}2\frac{5}{8}'' =$

7. 16-3¾″ − 4′-2½″ =

9. 8′-½″ − 3′-6⅝″ =

8. 1⅞″ − ¹³⁄₁₆″ =

10. 7′-¹³⁄₁₆″ − 4′-2½″ =

Math—Multiplying Fractions

1. ¾ × 8 =

6. 4½ × 2½ =

2. ⅜ × ½ =

7. ⅝ × ½ =

3. ⅞ × 2 =

8. 2¼ × 8¼ =

4. 2¼ × 13 =

9. 3¾ × 2 =

5. 3 × 1¹⁄₁₆ =

10. 1¼ × 2 =

Math—Dividing Fractions

1. $\frac{7}{8} \div 2 =$

2. $1\frac{1}{2} \div 3 =$

3. $4\frac{3}{4} \div 4 =$

4. $4\frac{1}{2} \div 2\frac{1}{4} =$

5. $\frac{3}{4} \div \frac{13}{16} =$

6. $\frac{13}{16} \div \frac{3}{4} =$

7. $6\frac{1}{2} \div 4 =$

8. $32 \div \frac{1}{2} =$

9. $\frac{7}{8} \div 1 =$

10. $1\frac{3}{4} \div 2 =$

Multiple Choice

_____ **1.** A wallboard hatchet is also known as a ___.
- A. wallboard hammer
- B. drywall hammer
- C. half-hatchet
- D. shingle hatchet

_____ **2.** Curved-claw hammers are commercially available in sizes ranging from ___ oz to ___ oz.
- A. 7; 20
- B. 8; 22
- C. 9; 24
- D. 10; 30

_____ **3.** A ___ is not a prying tool.
   A. ripping bar
   B. nail claw
   C. nail set
   D. flat bar

_____ **4.** A ___ is usually used to fasten floor underlayment.
   A. strike tacker
   B. hammer tacker
   C. gun tacker
   D. none of the above

_____ **5.** Standard screwdriver tips range in widths from ___″ to ___″.
   A. $\frac{1}{16}$; $\frac{1}{4}$
   B. $\frac{1}{8}$; $\frac{1}{4}$
   C. $\frac{1}{8}$; $\frac{3}{8}$
   D. $\frac{1}{8}$; $1\frac{1}{2}$

Printreading Symbols

Identify this symbol.

WC

_____

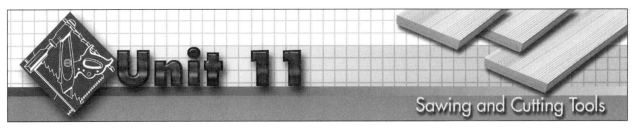
Name _____ Date _____

Multiple Choice

_____ **1.** The number printed on the blade of a handsaw indicates the ___.
   A. maximum depth of cut
   B. minimum depth of cut
   C. blade length
   D. number of teeth points per inch

_____ **2.** Hacksaws are used to cut ___.
   A. hardwood
   B. softwood
   C. metal
   D. none of the above

_____ **3.** Cold chisels are used for ___.
   A. cutting nails
   B. chipping concrete
   C. cutting through plaster
   D. all of the above

_____ **4.** Most handsaws have a ___ back.
   A. concave
   B. convex
   C. straight
   D. none of the above

_____ **5.** The sides of the blade of a good quality handsaw are ___.
   A. parallel
   B. thinner near the teeth
   C. thinner near the top
   D. thinner near the middle

_____ **6.** A utility knife has ___.
   A. fine teeth
   B. room for blade storage
   C. a 14″ blade
   D. all of the above

_____ **7.** Most ripsaws have a ___″ blade with ___ points per inch.
    A. 20; 4
    B. 22; 4½
    C. 24; 5
    D. 26; 5½

_____ **8.** Coping saws are generally used during ___.
    A. remodeling
    B. formwork
    C. finish work
    D. none of the above

_____ **9.** The teeth of crosscut saws are shaped like ___.
    A. knives
    B. chisels
    C. alternating knives and chisels
    D. none of the above

_____ **10.** Crosscut saws should be held at a ___° angle to the work.
    A. 30
    B. 45
    C. 60
    D. 75

Identification—Saws

_____ **1.** Hacksaw

_____ **2.** Compass saw

_____ **3.** Dovetail saw

_____ **4.** Coping saw

The Stanley Works

Ⓐ

The Stanley Works

Ⓑ

Ⓒ

Klein Tools, Inc.

Ⓓ

True-False

T   F   **1.** A saw cut is the width of the blade thickness.

T   F   **2.** The set of saw teeth helps prevent binding of the saw blade.

T   F   **3.** A backsaw is used with a miter box.

T   F   **4.** A butt chisel is used when mortising for door hinges.

T   F   **5.** An 11-point saw blade has smaller teeth than an 8-point saw blade.

T   F   **6.** Cold chisels are available with blade lengths from 2″ to 12″.

T   F   **7.** Crosscut saws for finish work normally have 10 to 12 points per inch.

T   F   **8.** Side-cutting pliers are used for pulling nails.

T   F   **9.** Compass saws are used to cut straight lines.

T   F   **10.** A keyhole saw, while similar to a compass saw, has fewer teeth per inch.

T   F   **11.** Knife blades should be adjusted to cut only through the designated material.

T   F   **12.** Tin snips may be used to trim metal studs.

T   F   **13.** Sawing with the grain requires more effort than sawing across the grain.

T   F   **14.** Flooring chisels are used primarily for finish work.

T   F   **15.** The printreading abbreviation for cedar is CE.

Math—Multiplication

1. $320 \times \frac{3}{8} =$

2. $720 \times 12 \times 4 =$

3. $400 \times \frac{5}{8} =$

4. $480 \times 1\frac{3}{4} =$

5. $\frac{3}{4} \times 1 \times \frac{3}{4} =$

6. $30 \times 8 \times \frac{1}{2} =$

7. ⅝ × 46 =

9. ⅛ × ¼ =

8. 479 × 329 =

10. 12 × 11 × ½ =

Math—Division

1. 4368 ÷ 56 =

6. 2 ÷ ½ =

2. 3.75 ÷ ⅓ =

7. 27,648 ÷ 432 =

3. 3.75 ÷ 3 =

8. 3½ ÷ ½ =

4. 9849 ÷ 21 =

9. 9 ÷ ⅜ =

5. 4½ ÷ 2¼ =

10. 480 ÷ ⅜ =

Boring and Clamping Tools

Name _____ Date _____

Completion

_____ 1. The ___ bit is available in lengths from 18″ to 24″.

_____ 2. The main parts of an auger bit are the tang, shank, spur, feed screw, and ___.

_____ 3. The sweep of a ratchet brace refers to the ___ of the circle made when turning the handle.

_____ 4. Auger bits for finish work have ___ screws.

_____ 5. Hand drills are normally used for drilling holes of ___″ diameter or less.

_____ 6. ___ bits are used to bore holes for cylindrical locks.

_____ 7. Ratchet braces can operate in a(n) ___ or a(n) ___ direction.

_____ 8. The number on the tang of an auger bit indicates size in ___ of an inch.

_____ 9. ___ bits are used to drive large screws.

_____ 10. ___ tools are used to hold and support materials.

_____ 11. The printreading abbreviation for ceiling is ___.

Identification—Auger Bit

_____ 1. Shank

_____ 2. Tang

_____ 3. Feed screw

_____ 4. Spur

_____ 5. Twist

Math—Addition

1. $\frac{3}{8} + \frac{3}{4} + 1\frac{1}{2} =$

2. $1\frac{1}{2} + 1\frac{1}{4} + \frac{5}{8} + \frac{1}{16} =$

3. $3\frac{1}{2} + \frac{5}{16} + \frac{7}{8} =$

4. $1\frac{13}{16} + 1\frac{5}{8} + \frac{3}{8} =$

5. $13\frac{13}{16} + 12\frac{5}{8} =$

6. $2\frac{7}{8} + 1\frac{3}{4} + 5\frac{5}{16} =$

7. $5\frac{3}{4} + 2\frac{1}{2} + 1\frac{9}{16} =$

8. $3\frac{5}{16} + 3\frac{5}{8} + 2\frac{7}{8} =$

9. $2\frac{5}{16} + 9\frac{3}{16} + 12\frac{1}{4} =$

10. $13\frac{15}{16} + 12\frac{9}{16} =$

Math—Subtraction

1. $13\frac{7}{8} - 1\frac{15}{16} =$

2. $22\frac{9}{16} - 3\frac{5}{8} =$

3. $134\frac{5}{8} - 111\frac{5}{16} =$

4. $19\frac{3}{4} - 3\frac{3}{8} =$

Smoothing Tools

Name _____ Date _____

True-False

T F **1.** Oil should not be used on a whetstone when sharpening a plane blade.

T F **2.** Jack planes are useful for all-purpose work.

T F **3.** Plane blades should be ground to an angle of 15° to 25°.

T F **4.** End grain should be planed from each edge.

T F **5.** A cabinet scraper is used for the final smoothing of a surface prior to sanding.

T F **6.** Rasps are used for fine finishing of edges.

T F **7.** The bullnose rabbet plane has an adjustable fence for setting the width of a cut.

T F **8.** When a plane iron is sharpened, the fine edge is obtained from a bench grinder.

T F **9.** Rasps produce a rough surface as wood is removed.

T F **10.** Serrated forming tools should not be used on plastic materials.

T F **11.** The printreading abbreviation for cement is CON.

Multiple Choice

_____ **1.** Serrated blades of forming tools ___.
   A. have hundreds of razor-sharp teeth
   B. do not clog easily
   C. cannot be sharpened
   D. all of the above

_____ **2.** Light scratches in veneers are removed with ___.
   A. scrapers
   B. block planes
   C. burnishers
   D. grinding wheels

_____ **3.** Rasps have ___-shaped teeth.
   A. knife
   B. chisel
   C. square
   D. triangular

_____ **4.** Two types of rabbet planes are ___ and ___.
   A. double edge; bullnose
   B. double edge; chamfer
   C. chamfer; duplex
   D. adjustable bullnose; bullnose

_____ **5.** Block planes are used for planing ___ and ___ surfaces.
   A. small; narrow
   B. small; wide
   C. long; narrow
   D. long; wide

_____ **6.** The burr on a scraper blade is made with a ___.
   A. file
   B. rasp
   C. burnishing tool
   D. none of the above

_____ **7.** A ___ plane is convenient for planing close to corners.
   A. jack
   B. bullnose
   C. block
   D. smooth

_____ **8.** The effectiveness of a plane is determined by the ___ and ___ of its plane iron.
   A. serration; weight
   B. size; length
   C. weight; size
   D. condition; sharpness

_____ **9.** ___ planes are normally used for fitting doors.
   A. Jointer
   B. Fore
   C. Block
   D. Jack

_____ **10.** Jointer planes are from ___" to ___" in length.
   A. 12; 16
   B. 16; 20
   C. 20; 24
   D. 24; 28

Completion

_____ **1.** The fine edge of a plane blade is produced on a(n) ___ after the blade has been ground.

_____ **2.** A block plane has a blade with a relatively ___ angle.

_____ **3.** When planing with a bench plane, downward pressure should be applied to the ___ at the beginning of the stroke.

_____ **4.** To obtain a smooth surface on a board, the direction of plane travel should be with the ___.

_____ **5.** The teeth of a rasp are ___-shaped.

Math

1. $\frac{3}{4} \times \frac{7}{8} =$

5. $.375 = \frac{}{8}$

2. $\begin{array}{r} 19.75 \\ \times \quad 22.50 \\ \hline \end{array}$

6. $\begin{array}{r} 3'\text{-}6\frac{1}{2}'' \\ + \quad 5'\text{-}3\frac{1}{4}'' \\ \hline \end{array}$

3. $56.5\overline{)272.33}$

7. $\begin{array}{r} 5'\text{-}10\frac{1}{4}'' \\ + \quad 6'\text{-}3\frac{1}{4}'' \\ \hline \end{array}$

4. $\dfrac{2 \times 12 \times 32}{4} =$

8. $\begin{array}{r} 14'\text{-}8\frac{3}{4}'' \\ + \quad 6'\text{-}3\frac{1}{2}'' \\ \hline \end{array}$

9. $\frac{3}{8} \times 1 \times \frac{5}{8} =$

13. $75 \times .75 =$

10. $\begin{array}{r} 27.52 \\ \times\ \ \ 3.15 \\ \hline \end{array}$

14. $\begin{array}{r} 8' \text{-} 0\frac{1}{2}'' \\ +\ \ 1' \text{-} 4\ \ '' \\ \hline \end{array}$

11. $933 \times \frac{1}{3} =$

15. $36 \div \frac{3}{4} =$

12. $\frac{1}{8} + \frac{3}{16} + \frac{1}{2} + \frac{3}{8} =$

Identification—Bench Plane

_____ **1.** Knob

_____ **2.** Adjusting nut

_____ **3.** Handle

_____ **4.** Toe

_____ **5.** Lateral adjusting lever

_____ **6.** "Y" adjusting lever

_____ **7.** Plane iron and plane iron cap

_____ **8.** Lever cap

_____ **9.** Cam

_____ **10.** Lever cap screw

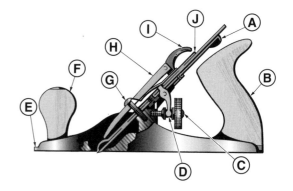

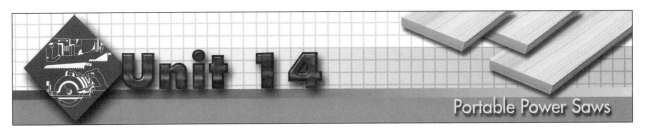

Unit 14

Portable Power Saws

Name _____ Date _____

Identification—Grounding System

_____ **1.** Grounded receptacle

_____ **2.** Power cord

_____ **3.** Grounding bar

_____ **4.** Copper ground wire

_____ **5.** Grounded plug

_____ **6.** Neutral conductor

_____ **7.** Electrical box

_____ **8.** Grounding screw

_____ **9.** Grounding slot

_____ **10.** Ground rod in earth

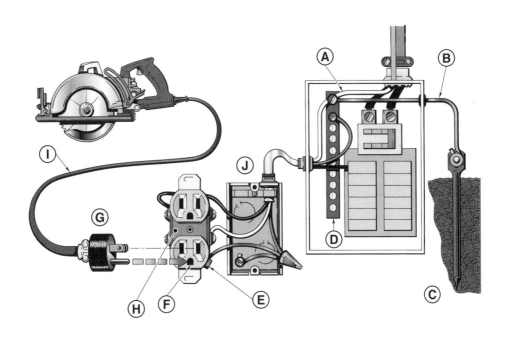

True-False

T   F   **1.** A ground fault circuit interrupter is used to provide electrical safety for workers.

T   F   **2.** Extreme caution should be exercised when using electric tools in damp locations.

T   F   **3.** A water pipe provides an excellent ground.

T   F   **4.** Cordless tools must be grounded while being used.

T   F   **5.** Double-insulated tools do not require a ground.

T   F   **6.** Circular electric handsaw blades cut from the underside of a board.

T   F   **7.** Circular electric handsaws may be used for ripping and crosscutting.

T   F   **8.** Electric tools should be disconnected when not in use.

T   F   **9.** Crosscut blades are used more often than combination blades.

T   F   **10.** Hollow-ground planer blades produce a smoother cut than crosscut or rip blades.

T   F   **11.** A circular electric handsaw should reach full speed before engaging the material to be cut.

T   F   **12.** Cuts on long pieces should be made between sawhorses.

T   F   **13.** A gasoline-operated saw should not be refueled when its engine is hot.

T   F   **14.** Saber saw blades cut on the upstroke.

T   F   **15.** Carbide-tipped blades retain their sharpness longer than conventional blades.

T   F   **16.** The printreading abbreviation for center is CENT.

Completion

_____   **1.** A ground fault circuit interrupter (GFCI) protects workers by opening a circuit in as little as ___ of a second.

_____   **2.** A(n) ___ GFCI should be inspected and tested before each use.

_____   **3.** Circular saws are equipped with blade guards ___ and below the base.

_____   **4.** ___ blades should be used when cutting masonry materials.

_____   **5.** Construction carpenters normally use circular electric handsaws with blades ranging from ___″ to ___″ in diameter.

_____   **6.** Electric tools having a conductor cord with a three-prong plug must be ___.

_____ **7.** The size of a circular electric handsaw is determined by the ___ of the largest blade it will accept.

_____ **8.** Circular electric handsaws are adjustable to cut angles from ___° to ___°.

_____ **9.** ___ blades are used for ripping and crosscutting.

_____ **10.** The major safety hazard when using electric tools is ___.

Short Answer

1. What safety precautions should be observed when working with electric tools in damp locations?

_____

_____

_____

_____

_____

_____

2. What are two methods of electric tool design that help prevent electric shock?

_____

_____

_____

_____

_____

3. What is the proper dress for a safety-conscious worker?

_____

_____

_____

_____

_____

Matching

_____ 1. Chisel-tooth combination blades

_____ 2. Side drive and worm drive

_____ 3. Aluminum oxide blades

_____ 4. Reciprocating saw

_____ 5. Saber saw

_____ 6. Chain saw

_____ 7. Jig saw

_____ 8. Grounded power tool

_____ 9. Crosscutting

_____ 10. Ripping

A. makes back and forth motion

B. used for cutting heavy timbers

C. types of circular electric saws

D. has three-conductor cord and three-prong plug

E. jig saw

F. used for cutting plastic laminates

G. used for cutting metals

H. cutting with the grain

I. cutting across the grain

J. makes circular motion

Math—Addition

1.
$$17'-6''$$
$$3'-2''$$
$$+\ 11'-9''$$

3.
$$14'-8\frac{1}{4}''$$
$$6'-5\frac{1}{2}''$$
$$+\ 2'-3\frac{1}{2}''$$

2.
$$3'-9\frac{3}{8}''$$
$$+\ 6'-4\frac{3}{4}''$$

4.
$$6'-8\frac{1}{2}''$$
$$5'-3\frac{1}{4}''$$
$$+\ 2'-2\frac{1}{8}''$$

Name _____ Date _____

Matching

_____ **1.** Miter saw

_____ **2.** Paper inserts

_____ **3.** Radial arm saw

_____ **4.** Miter gauge

_____ **5.** ⅛″ to ¼″

_____ **6.** 2″ × 12″

_____ **7.** 8″ to 20″

_____ **8.** 14″ to 16″

_____ **9.** Rip fence

_____ **10.** Push stick

A. cutoff saw

B. maximum table saw blade height above material

C. blade diameter range for radial arm saws

D. recommended sizes of radial arm saws for construction work

E. used for safely ripping wood

F. table saw accessory used for cross-cutting

G. table saw accessory used for ripping

H. portable saw used mainly for finish work

I. maximum stock size for frame-and-trim saw

J. used to increase width of dado cut

Multiple Choice

_____ **1.** The size of a radial arm saw is determined by the ___.
   A. horsepower rating of the saw
   B. length of the overarm
   C. largest blade it will accommodate
   D. size of the table

_____ **2.** A radial arm saw is also known as a(n) ___ saw.
   A. swing
   B. cutoff
   C. coping
   D. none of the above

_____ **3.** ___ can usually be prevented on a radial arm saw by tilting the saw table slightly back when the unit is set up.
A. Crawl
B. Ripping
C. Freehand cutting
D. none of the above

_____ **4.** In the interest of safety, a table saw blade should be no more than ___″ to ___″ above the material being cut.
A. ⅟₃₂; ⅟₁₆
B. ⅟₁₆; ⅛
C. ⅛; ¼
D. ¼; ¾

_____ **5.** ___ should be used when ripping on a table saw.
A. A rip fence
B. A push stick
C. Eye protection
D. all of the above

_____ **6.** ___ heads are used on a table saw to make rabbet and dado cuts.
A. Combination
B. Veneer
C. Dado
D. Rabbet

_____ **7.** A frame-and-trim saw can cut stock up to ___″ thick and ___″ wide.
A. 1; 6
B. 2; 6
C. 1; 12
D. 2; 12

_____ **8.** A frame-and-trim saw cannot make a ___.
A. rip
B. crosscut
C. miter
D. bevel

_____ **9.** Miter saws are used primarily for ___.
A. rough work
B. finish work
C. ripping
D. none of the above

_____ **10.** The printreading abbreviation for centerline is ___.
A. CENT
B. CL
C. CENL
D. CE

Identification—Miter Saw

_____ **1.** Angle adjustment knob

_____ **2.** Blade guard

_____ **3.** Blade

_____ **4.** Fence

_____ **5.** Miter Scale

_____ **6.** Table

_____ **7.** Motor

Math

_____ **1.** How many 1¾″ strips may be cut from a 10″ wide board? (Allow ⅛″ for each saw kerf.)

_____ **2.** A 10″ diameter saw blade has 60 teeth. How far apart are the teeth on the circumference of the blade?

_____ **3.** A 14′ board is cut into 12 equal lengths on a radial arm saw. How long is each piece? (Disregard saw kerf.)

_____ **4.** A table saw is used to cut a ¾″ wide rabbet in the center of a board 16″ long. What is the distance from the edge of the rabbet to the nearest end of the board?

_____ **5.** The following pieces of baseboard molding are required for a bedroom:
   two 12′-4″
   one 16′-0″
   one 2′-2″
   one 10′-4″
   What is the total number of feet of baseboard molding required?

Printreading Symbols

Identify this appliance.

_____

Math—Subtraction

1.   2060
     − 1981

7. $13\frac{7}{8} - 2\frac{1}{2} =$

2. $38'\text{-}6'' - 3'\text{-}4'' =$

8. $1'\text{-}9\frac{3}{4}'' - \frac{13}{16}'' =$

3. $13'\text{-}3'' - 5'\text{-}9'' =$

9. $13'\text{-}8'' - 6'\text{-}9'' =$

4. $16'\text{-}6\frac{5}{8}''$
     − $4'\text{-}2\frac{1}{4}''$

10. $3063 - 1311 =$

5. $4'\text{-}2\frac{1}{4}''$
     − $3'\text{-}8\frac{5}{8}''$

11. $\frac{3}{8} - \frac{3}{16} =$

6. $16'\text{-}8\frac{1}{8}''$
     − $7'\text{-}5\frac{1}{2}''$

12. $7'\text{-}6\frac{5}{16}'' - 1\frac{1}{4}'' =$

Name _____ Date _____

True-False

T   F   **1.** The size of a power drill is determined by the diameter of the largest bit shank that will fit into the drill chuck.

T   F   **2.** Solar-powered drills are available for use where there is no electricity.

T   F   **3.** Slower drill speeds should be used for harder materials.

T   F   **4.** D-handle power drills are generally used for heavy-duty work.

T   F   **5.** Variable-speed drills may be used to drive screws.

T   F   **6.** Corded drills are popular for high-torque applications.

T   F   **7.** Power drills are commercially available in sizes from ¼″ to 1½″.

T   F   **8.** Spade bits are also known as broad bits.

T   F   **9.** Straight-shank twist drills are commercially available in sizes from ¹⁄₆₄″ to ½″.

T   F   **10.** A drill is also referred to as a bit.

T   F   **11.** Carbide-tipped masonry drills are used to drill holes in brick, stone, and block.

T   F   **12.** Feeler drills are the most commonly used drills in the building trades.

T   F   **13.** Feeler bits can drill holes up to 24″ deep.

T   F   **14.** Ship auger bits without screwpoints are designed for cross-grain boring.

T   F   **15.** Hammer drills cannot be used to drill steel.

T   F   **16.** Hammer drills are used to drill large holes in masonry.

T   F   **17.** Rotary hammers can drill holes up to 8″ in diameter.

T   F   **18.** Hole saws can drill holes up to 6″ in diameter.

T   F   **19.** The 84° taper on the heads of flat-head screws allows recessing.

T   F   **20.** Hole saws are used to drill holes in glass.

T   F   **21.** Both hammer-drills and rotary hammers rotate and hammer during use.

T   F   **22.** The clutches of some hammer-drills can be disengaged so the tool can be used as a conventional drill.

T   F   **23.** Drywall screwdrivers recess the drywall surface when the screw is driven.

T   F   **24.** Drill bits should not be backed out of the material being drilled until the full hole depth has been reached.

T   F   **25.** A power drill should always be disconnected before removing or inserting a drill.

T   F   **26.** The printreading abbreviation for closet is CLOS.

Identification—Bits

_____   **1.** Hole saw

_____   **2.** Twist drill

_____   **3.** Ship auger bit

_____   **4.** Masonry bit

_____   **5.** Spade bit

_____   **6.** Countersink

Milwaukee Electric Tool Corp.
(A)

(D)

SCREWPOINT
Milwaukee Electric Tool Corp.
(B)

CARBIDE TIPS
(C)

Milwaukee Electric Tool Corp.
(E)

The Stanley Works
(F)

Identification—Power Drill

_____   **1.** Chuck

_____   **2.** Trigger switch

_____   **3.** Auxiliary handle

_____   **4.** Battery

_____   **5.** Torque adjustment

_____   **6.** Jaws

Skil Corporation

Short Answer

1. What is the proper grip and stance when using a power drill?

_____

_____

_____

2. What are the advantages and disadvantages of cordless drills?

_____

_____

_____

_____

3. Why have power drills replaced hand tools for drilling holes?

_____

_____

_____

_____

4. What are the differences between a hammer-drill and a rotary hammer?

_____

_____

_____

_____

_____

5. What is the advantage of reduced-shank drills over straight-shank drills?

_____

_____

_____

Math—Multiplication

1. 62.05
  × 1.75

9.   780
  × 1.75

2. 32.6 × .75 =

10. 20 × 12 × ¾ =

3. 4½ × 12 =

11. 16 × 2 × ½ =

4. ⅞ × 1½ =

12.   368
  × 12.25

5. ¾ × 9 =

13. ⅜ × 15 × 12 =

6. 20 × 16.75 =

14. 9½ × 11 =

7. .375 × .55 =

15. 6 × 18 × ¼ =

8.   3.14
  × 12

Unit 17

Portable Power Planes, Routers, and Sanders

Name _____ Date _____

Completion

_____ 1. The exposed ___ of a portable power plane has the greatest potential for danger to the user.

_____ 2. Routers are used for ___ and ___ operations.

_____ 3. Heavy-duty belt sanders use belts from ___″ to ___″ long.

_____ 4. Adjustable fences on portable power planes allow planing bevels up to ___°.

_____ 5. The plug should be disconnected when changing router bits and making ___.

_____ 6. A(n) ___ finish sander has a circular and oscillating motion.

_____ 7. Cutters on portable power planes can be as wide as ___″.

_____ 8. A laminate trimmer is used with a router to cut ___.

_____ 9. A portable power plane should not be set down until the ___ has stopped.

_____ 10. ___ should be worn including the proper eye, hearing, and respiratory protection.

_____ 11. Four types of material from which sandpaper is made are ___, ___, ___, and ___.

_____ 12. Portable power ___ planes can be used to plane end grain.

_____ 13. The three grades of abrasive sandpapers are ___, ___, and ___.

_____ 14. The exposed ___ represents the most potential danger when using a router.

_____ 15. The change of grain direction on the edge of a door is known as ___.

_____ 16. ___ and ___ are natural minerals from which sandpaper is made.

_____ 17. ___ sanders are used for finish sanding.

_____ 18. Grit refers to the number of ___ particles per square inch.

_____ 19. Three types of finish sanders are ___, ___, and ___.

_____ 20. ___ and ___ sandpapers are recommended for use with electric sanders.

_____ 21. ___ sanders are used for heavy sanding.

_____ 22. The printreading abbreviation for column is ___.

Identification—Portable Power Plane

_____ **1.** Front shoe

_____ **2.** Depth adjustment

_____ **3.** Front handle

_____ **4.** Motor housing

_____ **5.** Trigger switch

_____ **6.** Rear shoe

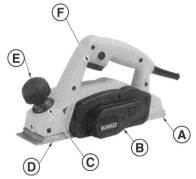

DeWALT Industrial Tool Co.

Identification—Router Bits

_____ **1.** Bead   _____ **6.** Dovetail

_____ **2.** Corner round   _____ **7.** Panel pilot

_____ **3.** Cove   _____ **8.** 45° bevel chamfer

_____ **4.** Sash cope   _____ **9.** Sash bead

_____ **5.** Core box   _____ **10.** Rabbeting

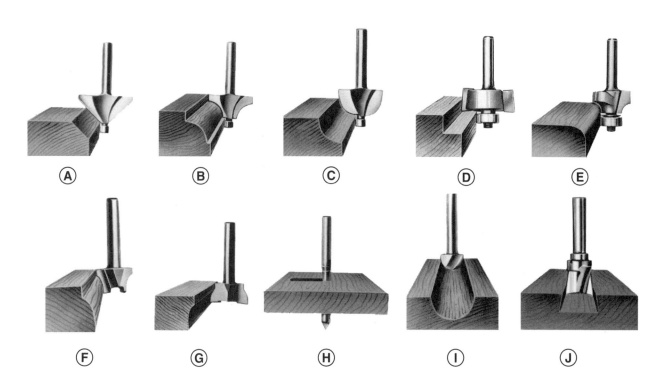

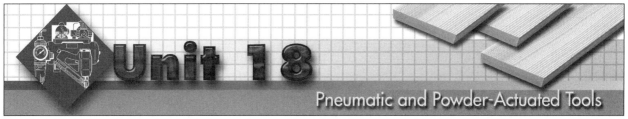

Name _____ Date _____

True-False

T   F   **1.** Framing nailers are commonly used by carpenters.

T   F   **2.** Drive pins are used with pneumatic nailers.

T   F   **3.** A nail will not split the wood as easily as a staple when driven near the end of a board.

T   F   **4.** Powder-actuated fastening tools are used primarily to fasten wooden materials to wooden joists and studs.

T   F   **5.** Penetration depths of powder-actuated tools may be controlled by the load of the cartridge used.

T   F   **6.** Air pressure for pneumatic tools is adjusted by regulators.

T   F   **7.** Nails for pneumatic nailers are available in strips and coils.

T   F   **8.** Pneumatic staplers can accommodate more than one size of staple.

T   F   **9.** The power level for powder loads has a scale from 1 to 10.

T   F   **10.** The printreading abbreviation for flashing is FL.

Completion

_____ **1.** Pneumatically powered tools are driven by ___ air.

_____ **2.** Some pneumatic nailers can be adjusted to ___ nails below the surface.

_____ **3.** A(n) ___ or a(n) ___ is the fastening device fired from powder-actuated tools.

_____ **4.** Concrete thickness should be at least ___ times the fastener shank penetration.

_____ **5.** Pneumatically driven nails have ___ holding power than hammer-driven nails.

_____ **6.** Air ___ required for a job depends on the size and type of operation being performed.

_____ **7.** Drive pins driven into steel often have a(n) ___ shank for extra holding power.

_____ **8.** Powder loads are ___ for ease of identification.

_____ **9.** Single-shot powder-actuated tools must be ___ loaded each time the tool is discharged.

_____ **10.** Compressor air hoses that are more than ___″ in diameter must have a safety device to stop air flow in case of a hose failure.

Identification—Pneumatic Nailer

_____ **1.** Chamber

_____ **2.** Valve plunger

_____ **3.** Air reservoir

_____ **4.** Trigger

_____ **5.** Magazine

_____ **6.** Piston

Short Answer

1. What is the major safety feature of all pneumatic nailers and staplers?

_____

_____

2. Discuss how to check a plaster-covered wall to verify that the wall is made of concrete before firing a nail or stud into the wall.

_____

_____

Math

1. $6'\text{-}3\frac{1}{2}'' + 4'\text{-}9'' + 5'\text{-}1'' =$

9. $1\frac{3}{8} + 9\frac{1}{2} + 5\frac{1}{4} =$

2. $21.75\overline{)852.6}$

10. $5.75 \times 12.5 =$

3. $\frac{3}{8} \times 3\frac{1}{2} =$

11.
$$\begin{array}{r} 52.50 \\ -43.97 \\ \hline \end{array}$$

4. $\frac{5}{16} \times 2 =$

12. $98 \times \frac{3}{8} =$

5. $.375 \times 4 =$

13. $2'\text{-}9'' + 3'\text{-}1'' + 6'\text{-}10'' =$

6. $4\overline{)3 \times 24}$

14.
$$\begin{array}{r} 12'\text{-}6\frac{1}{4}'' \\ -9'\text{-}7\frac{1}{2}'' \\ \hline \end{array}$$

7. $4 \times 4\overline{)12 \times 16}$

15. $1 \times 1 \times 12\overline{)24 \times 12 \times 2}$

8. $2.5\overline{)5.5}$

Name _____ Date _____

Matching

_____ **1.** Oxygen cylinder   **A.** welding stick

_____ **2.** Acetylene cylinder   **B.** threads found on oxygen equipment

_____ **3.** Flammable material   **C.** used to hold oxygen under pressure

_____ **4.** Acetylene hose   **D.** tip

_____ **5.** Oxygen hose   **E.** jumps between electrode and workpiece

_____ **6.** Electrode   **F.** red hose

_____ **7.** Nozzle   **G.** used to hold acetylene under pressure

_____ **8.** Right-handed threads   **H.** threads found on acetylene equipment

_____ **9.** Left-handed threads   **I.** green hose

_____ **10.** Arc   **J.** material that burns easily

Completion

_____ **1.** Carpenters repair steel forms and weld ___ bars and structural members of ___ walls.

_____ **2.** Carpenters must be ___ by proper authorities before performing welding or metal-cutting operations.

_____ **3.** ___ equipment is used primarily for cutting steel in the building trades.

_____ **4.** Metal is fused together by the heat of an electric arc in ___ welding.

_____ **5.** ___ welding is used more in the building trades than ___ welding.

_____ **6.** Current for electric arc welding machines may be ___, ___, or ___.

_____ **7.** ___ conduct electrical current from the welding machine, through the electrode holder, to the work.

_____ **8.** Electrode holders are ___ to prevent electrical shock.

_____ **9.** ___ should be worn when welding metals that may give off toxic fumes.

_____ **10.** ___ and ___ rays from electric arc welding can cause severe damage to the eyes.

_____ **11.** A(n) ___ is attached to the work in electric arc welding.

_____ **12.** Metal is fused together by the heat of a welding flame in ___ welding.

_____ **13.** ___ equipment can be used for welding or cutting.

_____ **14.** ___ on oxygen and acetylene tanks are used to adjust pressure.

_____ **15.** ___ parts of an electrode holder or gun must be fully insulated.

_____ **16.** Welding cables must be free of ___ or other defects.

_____ **17.** A power ___ switch should be located near the electric arc welding machine.

_____ **18.** Oxygen and acetylene lines should be ___ before the torch is lit.

_____ **19.** Metal being welded must reach its ___ point before fusion can occur.

_____ **20.** ___ on oxygen and acetylene cylinders should always be opened slowly.

True-False

T   F   **1.** Carpenters may weld straps and hangers in place.

T   F   **2.** Welding courses may be offered as part of an apprenticeship program.

T   F   **3.** Threads of nuts and couplings on acetylene lines are always right-handed.

T   F   **4.** Oxyacetylene flames may reach temperatures of 16,800°F.

T   F   **5.** Oxygen should not be used to dust off clothing.

T   F   **6.** Oxygen should never be used as a substitute for compressed air.

T   F   **7.** Stick electrodes must be replaced when they are down to 3″ in length.

T   F   **8.** An acetylene hose is red.

T   F   **9.** Threads of nuts and couplings on oxygen lines may be right-handed or left-handed.

T   F   **10.** Oxygen hoses may be green or blue.

T   F   **11.** The major difference between cutting and welding with oxyacetylene is the gas mixture.

T   F   **12.** Oxygen and acetylene flow through a common hose from the tanks to the torch.

T   F   **13.** Electric arc welding utilizes an oxygen-acetylene fuel mixture to fuse metals together.

T   F   **14.** The arc of electric arc welding should never be viewed with the naked eye.

T   F   **15.** Oxygen and acetylene tanks are portable and may be moved about on the job site, provided proper transportation means are available.

T   F   **16.** The printreading abbreviation for concrete is CONC.

Identification—Oxyacetylene Equipment Setup

_____ **1.** Oxygen needle valve

_____ **2.** Torch

_____ **3.** Acetylene pressure adjustment screw

_____ **4.** Acetylene cylinder high-pressure gauge

_____ **5.** Welding tip

_____ **6.** Oxygen pressure adjustment screw

_____ **7.** Oxygen torch working low-pressure gauge

_____ **8.** Acetylene torch working low-pressure gauge

_____ **9.** Oxygen cylinder high-pressure gauge

_____ **10.** Oxygen cylinder valve

_____ **11.** Acetylene cylinder

_____ **12.** Hoses

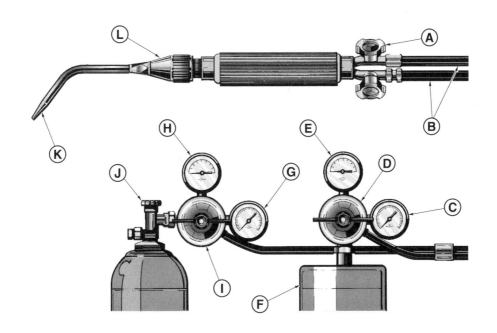

Math

1. $4\frac{7}{8} \times 5 =$

2. $5\frac{7}{16} - 2\frac{3}{8} =$

3. $5\frac{3}{16} + 2\frac{9}{16} =$

4. $19.357 - 12.60 =$

5. $12'\text{-}4\frac{1}{8}'' + 2'\text{-}9\frac{1}{16}'' =$

6. $12'\text{-}6\frac{1}{2}'' - 9'\text{-}2'' =$

7. $378 \overline{)9828}$

8. $1\frac{1}{2} \div 3 =$

9. $1\frac{1}{2} \times 3 =$

10. $\frac{3}{4}$ of $120 =$

11. $\frac{2}{3}$ of $360 =$

12. $1 \times 12 \times 12 \overline{)1 \times 6 \times 96}$

13. $2 \times 4 \times 12 \overline{)1 \times 8 \times 12}$

14. $1 \times 12 \times 12 \overline{)2 \times 10 \times 144}$

15. $1 \times 12 \times 12 \overline{)1 \times 3 \times 48}$

Unit 20

Scaffolds, Aerial Lifts, and Ladders

Name _____ Date _____

True-False

T   F   **1.** A scaffold is a permanent work platform.

T   F   **2.** The maximum intended load is the total of all loads on a scaffold.

T   F   **3.** Lumber with loose knots may be used for scaffolds.

T   F   **4.** All scaffolds more than 10′ above the ground must have a guardrail system.

T   F   **5.** Toeboards are installed to protect workers' toes.

T   F   **6.** Wooden or metal brackets may be used for carpenter's scaffolds.

T   F   **7.** Sectional metal-framed scaffolds are the primary type of scaffold used in construction.

T   F   **8.** A metal scaffold may be erected within 8′ of noninsulated power lines.

T   F   **9.** Hard hats and fall-arrest equipment are optional when working on elevated work platforms or buckets of aerial lifts.

T   F   **10.** The printreading abbreviation for concrete block is CB.

Multiple Choice

_____   **1.** When tying ends of two ropes of different thicknesses, a ___ knot should be used.
   A. sheet bend
   B. bowline
   C. square
   D. four-in-one

_____   **2.** A putlog is a ___ truss that extends between two separate scaffolds.
   A. horizontal
   B. vertical
   C. diagonal
   D. all of the above

_____   **3.** Scaffold planks must extend ___″ to ___″ past the end supports.
A. 4; 10
B. 6; 12
C. 10; 18
D. 12; 24

_____   **4.** Tie-ins and ___ provide support for a scaffold and prevent it from tipping.
A. guard rails
B. platforms
C. guylines
D. sills

_____   **5.** As a general rule, scaffolds must support ___ times the weight to which they will be exposed.
A. two
B. four
C. six
D. eight

_____   **6.** Safety nets must be used for work ___′ or more above the ground when the worker is not otherwise protected.
A. 8
B. 12
C. 15
D. 25

_____   **7.** Suspension scaffolds are supported by wire ropes that must support at least ___ times the maximum intended load.
A. four
B. six
C. eight
D. ten

_____   **8.** The distance from the work platform to the upper surface of the top rail should be ___″ to ___″.
A. 32; 48
B. 36; 42
C. 36; 48
D. 38; 45

_____   **9.** Platform planks for scaffolds should overlap a minimum of ___″.
A. 12
B. 18
C. 24
D. 30

_____   **10.** When tying ends of two ropes of the same thickness, a ___ knot should be used.
A. single-sheet bend
B. bowline
C. square
D. four-in-one

Completion

_____ 1. Three basic types of sectional metal-framed scaffolds are welded-frame, ___, and system scaffolds.

_____ 2. A carpenter's sawhorse is normally ___″ high.

_____ 3. Tube-and-clamp scaffolds allow more ___ in the shape of the scaffold so that the scaffold can be built around circular structures.

_____ 4. Mobile scaffolds require a(n) ___ paved or concrete surface.

_____ 5. Commercially available ladders are made of ___, ___, and ___.

_____ 6. The horizontal distance between the base of a ladder and the structure should be ___ the distance from the ground to the top support of the ladder.

_____ 7. The ___ hitch is used for tying off members to be raised or lowered.

_____ 8. ___ ladders conduct electricity when wet.

_____ 9. Pump jacks may be used for heights of up to ___′.

_____ 10. Uprights for pump jacks should be a minimum of ___″ square.

Identification—Ladders

_____ 1. Stepladder

_____ 2. Extension ladder

_____ 3. Single ladder

_____ 4. Fixed ladder

Ballymore Company, Inc.

Ⓐ

Werner Ladder Co.

Ⓑ

Ⓒ

Ⓓ

Identification—Tube-and-Clamp Scaffold

_____ **1.** Guardrail

_____ **2.** Horizontal brace

_____ **3.** Diagonal brace

_____ **4.** Platform

_____ **5.** Toeboard

_____ **6.** Midrail

_____ **7.** Cross brace

_____ **8.** Post or vertical leg

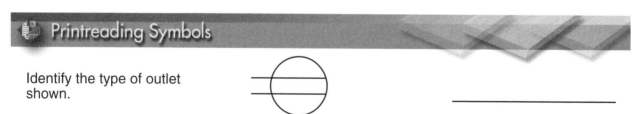

Identify the type of outlet shown.

_____

Name _____   Date _____

Matching

_____ **1.** Backhoe loader

_____ **2.** Aerial lift

_____ **3.** Telescopic-boom crane

_____ **4.** Lattice-boom crane

_____ **5.** Rough-terrain forklift

_____ **6.** Z-boom lift

_____ **7.** Material lift

_____ **8.** Tower cranes

A. used for digging and loading operations

B. used on heavy construction projects

C. has two or more hinged sections

D. piece of extendable and/or articulating equipment

E. traverses terrain of construction sites

F. places prefabricated panels on small jobs

G. has telescopic arms

H. has gridwork of steel reinforcing members

Completion

_____ **1.** Carpenters may set lines and lay out areas for ___ and ___ with earth-moving machinery.

_____ **2.** A(n) ___ is used to strip rocks and topsoil from the excavation site.

_____ **3.** A(n) ___ is used for final grading operations on large construction sites.

_____ **4.** Inside-climbing tower cranes are normally raised as every ___ to ___ stories of a high-rise structure are completed.

_____ **5.** As the height of a high-rise structure increases, an inside-climbing tower crane is elevated with ___.

_____ **6.** Backhoe loaders are equipped with ___ to provide stability during trenching operations.

_____ **7.** ___ signals are used to direct crane operators of mobile and stationary cranes.

_____ **8.** A(n) ___ is used primarily to pick up and deposit loose soil and rocks into trucks for removal.

_____ **9.** The two main types of tower cranes used on heavy construction projects are ___ and ___.

_____ **10.** A(n) ___ is used to remove soil and deposit it into trucks.

_____ **11.** ___ are commonly used in road construction to ensure the proper roadbed grade.

_____ **12.** The ___ of a grader can be adjusted to various angles and positions.

_____ **13.** A(n) ___ must be set up on the roof of a high-rise structure to remove the climbing crane.

_____ **14.** Climbing cranes are dismantled into ___ before they are lowered to the ground.

_____ **15.** A(n) ___ is used to bore holes for concrete picrs or piles.

Identification—Hand Signals

_____ **1.** Dog everything

_____ **2.** Raise boom

_____ **3.** Hoist

_____ **4.** Emergency stop

_____ **5.** Travel

_____ **6.** Lower

_____ **7.** Extend boom

_____ **8.** Lower boom

_____ **9.** Move slowly

_____ **10.** Retract boom

True-False

T   F   **1.** The amount of excavation on a job site depends on the size of the structure to be erected.

T   F   **2.** A bulldozer blade is normally mounted parallel to the line of travel.

T   F   **3.** Mobile and stationary cranes are used on construction projects.

T   F   **4.** An excavator may be used for loading operations.

T   F   **5.** An inside-climbing tower crane may be set up in an elevator shaft.

T   F   **6.** Free-standing tower cranes must be bolted to a concrete pad.

T   F   **7.** A derrick has a swinging mast and a vertical boom.

T   F   **8.** When working beneath a crane, workers should always walk in the crane's line of travel.

T   F   **9.** Hand signals may be given by any individual allowed on the job site.

T   F   **10.** Workers may ride on a load only if it is attached to the cable of a crane.

T   F   **11.** The printreading abbreviation for cornice is COR.

Identification—Mobile Lattice-Boom Cranes

_____ **1.** Wheel-mounted crane

_____ **2.** Crawler crane

_____ **3.** Locomotive crane

_____ **4.** Motor truck crane

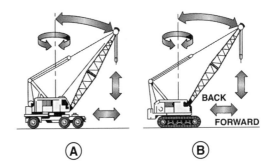

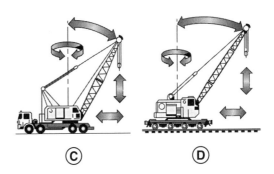

Printreading Symbols

Identify the type of wall shown.

_____

Name _____ Date _____

Completion

_____ 1. Slips and falls from elevated work surfaces and ladders account for ___ of all construction injuries.

_____ 2. Sheet piling for trenches over 8′ deep must be at least ___″ thick.

_____ 3. Guardrails should be nailed across all wall openings from which there is a drop of more than ___′.

_____ 4. Heavy-duty safety shoes with reinforced ___ toes should be worn to protect a worker's feet.

_____ 5. Lumber piles that are to be handled with equipment should not exceed ___′ height.

_____ 6. The banks of an excavation in average soils should slope ___°.

_____ 7. Sheet piling for excavation walls is positioned by cranes and driven into the soil with a pile-driving ___.

_____ 8. Walls of excavations should be ___ or ___ to prevent their banks from collapsing.

_____ 9. Lumber piles that are to be handled manually should not exceed ___′ in height.

_____ 10. Areas within ___′ of a building should be reasonably level.

_____ 11. ___ consists of interlocking steel sections used as walls in shoring operations.

_____ 12. ___ are erected around construction sites to prevent access by unauthorized personnel.

_____ 13. ___ within 5′ of one another may be used for shoring in hard, compact soil.

_____ 14. ___ must be constructed around floor openings that are framed for stairwells.

_____ 15. Work surfaces ___′ or higher should be guarded by guardrails.

_____ 16. Class ___ fires may occur with live electrical equipment.

_____ 17. Class ___ fires are extinguishable with water.

_____ **18.** ___ materials can create flammable vapors by evaporating at normal temperatures and pressures.

_____ **19.** Fire extinguishers must be located within ___' of all work areas.

_____ **20.** Class ___ fires may occur with flammable liquids.

Identification—Fire Extinguisher Classes

_____ **1.** Combustible metals

_____ **2.** Electrical equipment

_____ **3.** Flammable liquids

_____ **4.** Ordinary combustibles

_____ **5.** Commercial cooking grease

TRASH • WOOD • PAPER
WOOD SCRAPS
A
Ⓐ

LIQUIDS • GREASE
SOLVENT CEMENT
B
Ⓑ

MOTORS • TRANSFORMERS
MOTOR
C
Ⓒ

ZIRCONIUM • TITANIUM
D
METAL
D
Ⓓ

GREASE
DEEP FAT FRYER
Ⓔ

True-False

T   F   **1.** Strain or overexertion is the most common injury suffered by construction workers.

T   F   **2.** Slopes for excavation walls must meet a ratio of ¾ vertical to 1 horizontal.

T   F   **3.** The most common cause of death from job site accidents involves moving vehicles.

T   F   **4.** Interlocking sheet piling is not reusable.

T   F   **5.** Power-driven buggies may not be used on job-constructed ramps.

T   F   **6.** Class A and Class B fires should be extinguished with water.

T   F   **7.** OSHA lists safety procedures designed to protect workers on the job.

T   F   **8.** A wood-burning fire is a Class A fire.

T   F   **9.** The noise level of certain operations on construction jobs can cause permanent ear damage over a period of time.

T   F   **10.** Respiratory protection is required when workers are exposed to airborne hazards.

T   F   **11.** The printreading abbreviation for corrugated is CORR.

Short Answer

1. Discuss precautions carpenters can take to help reduce the possibility of accidents.

_____

_____

_____

_____

2. Discuss proper work habits carpenters should follow to protect themselves and their fellow workers.

_____

_____

_____

_____

3. What is the proper procedure for lifting heavy objects?

_____

_____

_____

4. What is the proper procedure a worker should follow if a fire occurs on the job site?

_____

_____

_____

5. Discuss proper procedures for storing flammable materials.

_____

_____

_____

Unit 23

Building Design, Plans, and Specifications

Name _____ Date _____

True-False

T   F   **1.** Pueblo revival architecture is commonly used in New England.

T   F   **2.** Houses built on speculation must be sold prior to construction.

T   F   **3.** Sun and wind factors may affect the orientation of a building.

T   F   **4.** One-and-one-half story houses have high-pitched roofs.

T   F   **5.** A split-level house always has an even number of floors.

T   F   **6.** Soil conditions determine the type of foundation that can be used.

T   F   **7.** A habitable room is a room used for living purposes.

T   F   **8.** Custom-built houses are designed and built to meet the wishes of a particular owner.

T   F   **9.** Stock plans are developed by the builder for a specific customer.

T   F   **10.** Specifications are commonly divided into divisions that pertain to different work areas on the project.

T   F   **11.** The printreading symbol for detail is DT.

Completion

_____   **1.** The exterior appearance of most buildings is either ___ or contemporary.

_____   **2.** ___ analyze the ground conditions of a building site.

_____   **3.** ___ refers to the position of a building on the lot.

_____   **4.** One-family dwellings are normally built over a full-basement, crawl-space, or ___ foundation system.

_____   **5.** Four basic shapes of one-family dwellings are ___, ___, ___, and ___.

_____   **6.** Four basic types of one-family dwellings are ___, ___, ___, and ___.

_____   **7.** ___ design buildings and supervise the drawing of plans.

_____ **8.** Newly constructed buildings must conform to the ___ code and ___ regulations for the area.

_____ **9.** ___ design the load-bearing portions of large buildings.

_____ **10.** ___ are legal documents that help clarify working drawings.

Identification—Roofs

_____ **1.** L-shape

_____ **2.** Rectangular

_____ **3.** T-shape

_____ **4.** U-shape

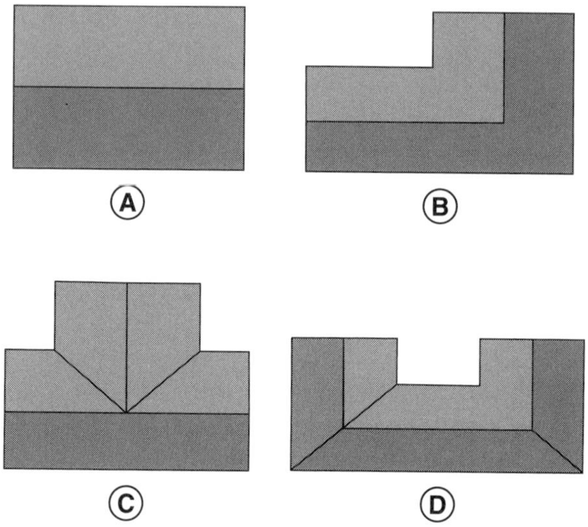

Ⓐ

Ⓑ

Ⓒ

Ⓓ

Name _____ Date _____

Matching

_____ **1.** Diameter

_____ **2.** Bedroom

_____ **3.** Elevation

_____ **4.** South

_____ **5.** Drain

_____ **6.** Steel

_____ **7.** Electric

_____ **8.** Wood

_____ **9.** Gauge

_____ **10.** Interior

A. GA

B. STL

C. S

D. ELEC

E. DIAM

F. WD

G. EL

H. DR

I. INT

J. BR

Identification—Lines Used for Drawings

_____ **1.** Object line

_____ **2.** Leader line

_____ **3.** Centerline

_____ **4.** Dimension line

_____ **5.** Extension line

_____ **6.** Break line

_____ **7.** Cutting plane line

_____ **8.** Hidden line

Identification—Measurement

The ¼″ = 1′-0″ scale on page 328 may be used. An architect's scale may also be used.

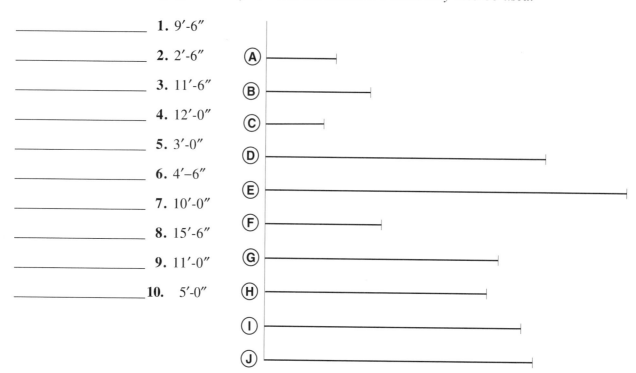

_____ **1.** 9′-6″

_____ **2.** 2′-6″ (A)

_____ **3.** 11′-6″ (B)

_____ **4.** 12′-0″ (C)

_____ **5.** 3′-0″ (D)

_____ **6.** 4′–6″ (E)

_____ **7.** 10′-0″ (F)

_____ **8.** 15′-6″ (G)

_____ **9.** 11′-0″ (H)

_____ **10.** 5′-0″ (I)

(J)

True-False

T   F   **1.** A cutting plane line allows interior features of an object to be viewed.

T   F   **2.** An orthographic view of a building shows the building from above and from each side.

T   F   **3.** All plans on a print relate to each other.

T   F   **4.** The same scale must be used on all drawings of a set of prints.

T   F   **5.** A scale of ¼″ = 1′-0″ is the most commonly used scale on a set of architectural prints.

T   F   **6.** Prints give a pictorial view of each part of a building.

T   F   **7.** An engineer's scale is used for scaling dimensions on architectural prints.

T   F   **8.** Written notes on architectural plans are abbreviated whenever possible.

T   F   **9.** Plumbing fixtures on a set of prints are identified by pictorial drawings.

T   F   **10.** Architectural plan symbols may represent materials, fixtures, or structural parts.

Identification—Symbols

_____ 1. Three-way switch

_____ 2. Duplex receptacle outlet

_____ 3. Air return

_____ 4. Thermostat

_____ 5. Concrete

_____ 6. Cavity wall

_____ 7. Outlet box and incandescent lighting fixture

_____ 8. Motor

_____ 9. Switch control of light

_____ 10. Junction box

_____ 11. Frame wall

_____ 12. Firebrick

_____ 13. Single-pole switch

_____ 14. Range outlet

_____ 15. Smoke detector

_____ 16. Fluorescent lighting fixture

_____ 17. Duct

_____ 18. Water heater

_____ 19. Water closet

_____ 20. Range top

_____ 21. Fan

_____ 22. Lighting panel

_____ 23. Recessed outlet box and incandescent lighting fixture

_____ 24. Refrigerator

_____ 25. Radiator

(A) T

(B) ◯ OR ◯

(C) ▭

(D) WC

(E) ⊙⊙ ⊙⊙

(F) S₃

(G) RAD

(H) ▭○

(I) ▭

(J) WH

(K) REF

(L) ⊖

(M) M

(N) ▢ OR ⊞

(O) ▱

(P) J

(Q) ⊠

(R) SD

(S) ▮

(T) S

(U) ▱

(V) F OR ∞

(W) S ⌒ ◯

(X) ▭

(Y) ⊖ R

Completion

_____ **1.** Plans are drawn to a specific ___ as indicated on the prints.

_____ **2.** Print drawings show ___ views rather than pictorial views of each part of the building.

_____ **3.** A(n) ___ line is used when interior features must be shown.

_____ **4.** Pictorials or abbreviations that represent a material, fixture, or structural part are known as ___.

_____ **5.** A(n) ___ line indicates a shortened view of a part that has a uniform shape.

_____ **6.** ___ lines are terminated by an arrowhead or dot on each end.

_____ **7.** A(n) ___ line points from a note or measurement to a particular part of the plan.

_____ **8.** Dashed lines known as ___ lines are used to show the unseen edges of surfaces.

_____ **9.** The symbol ___ on an electrical plan indicates a four-way switch.

_____ **10.** The abbreviation for solar panel is ___.

_____ **11.** The abbreviation SEW on a set of plans indicates a(n) ___.

_____ **12.** The symbol for ___ is a rectangular box containing numerous, cross-hatched diagonal lines.

_____ **13.** ___ are measurements that give the distances between different points such as walls, columns, beams, and other structural parts.

_____ **14.** A symbol drawn as a square with the letters FD immediately below represents a(n) ___.

_____ **15.** The abbreviation for jamb is ___.

Name _____ Date _____

Multiple Choice

_____ **1.** Information given on a plot plan includes the ___.
  A. exact location of the building on the property
  B. high and low points of the property
  C. size and shape of the building
  D. all of the above

_____ **2.** Retaining walls may be constructed with ___.
  A. concrete or concrete blocks
  B. railroad crossties
  C. rock piles
  D. all of the above

_____ **3.** A plot plan does not include ___.
  A. elevations
  B. swales
  C. driveway locations
  D. number of windows

_____ **4.** Easements on plot plans may indicate provisions for ___.
  A. planting trees
  B. use by public utility companies
  C. front setbacks
  D. all of the above

_____ **5.** The term *elevation*, when used in conjunction with a plot plan, refers to ___.
  A. grade
  B. heights established for different levels of a building
  C. slope of the swale
  D. all of the above

_____ **6.** Finish grades and elevations on the plot plan are normally determined from data provided by ___ or engineers.
  A. property owners
  B. contractors
  C. surveyors
  D. carpenters

_____ 7. All finish grades on a plot plan are based upon their relation to the ___.
    A. property lines
    B. existing street
    C. benchmark
    D. none of the above

_____ 8. Benchmarks may be identified by a ___.
    A. plugged pipe driven into the ground
    B. brass marker
    C. wood stake
    D. all of the above

_____ 9. The plot plan shows finish grades for ___.
    A. all corners of the lot
    B. all building corners
    C. the driveway
    D. all of the above

_____ 10. Bench marks are established by a ___.
    A. surveyor
    B. contractor
    C. carpenter
    D. property owner

Completion

_____ 1. The different sides of a building may be referred to by ___ directions.

_____ 2. A benchmark grade figure is either the number of feet above sea level or it is the number ___.

_____ 3. A(n) ___ refers to the grade for channeling water away from the building.

_____ 4. ___ walls are used to prevent earth from sliding.

_____ 5. The outline of the building to be constructed on a lot is shown by ___ lines on the plot plan.

_____ 6. The property line on a plot plan shows the ___ of a lot.

_____ 7. Sloped lots may require ___ to provide water runoff away from the building.

_____ 8. A benchmark is also known as a job ___.

_____ 9. ___ lines on a plot plan may show the existing and/or finish grades.

_____ 10. Grades on a plot plan are generally given in ___ and ___ of a foot.

Identification—Plot Plan

_____ **1.** Property lines

_____ **2.** Building lines

_____ **3.** Electrical utilities

_____ **4.** Front setback

_____ **5.** Walk

_____ **6.** Planter strip

_____ **7.** Terrace

_____ **8.** Finish floor elevation

_____ **9.** Benchmark

_____ **10.** Finish grade

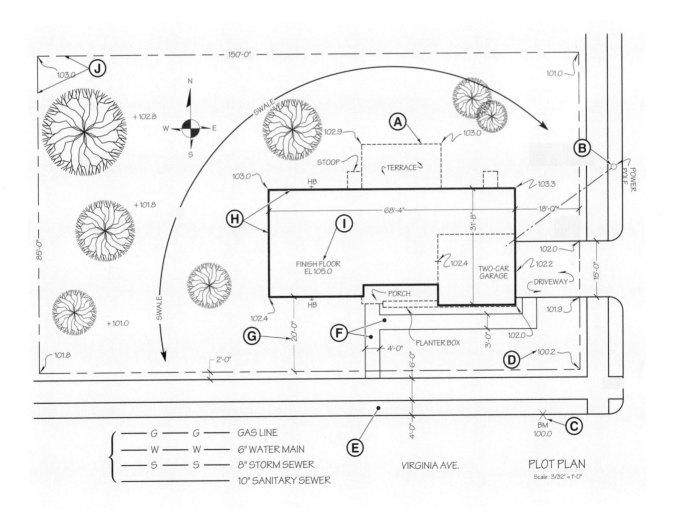

True-False

T	F	**1.** A plot plan is also known as a site plan.
T	F	**2.** A plot plan usually includes an arrow designating North.
T	F	**3.** A property line on a plot plan is also known as a lot line.
T	F	**4.** Widths of sidewalks and driveways may be found on the plot plan.
T	F	**5.** Finish floor elevations are not given on the plot plan.
T	F	**6.** The location of public utilities is not given on the plot plan.
T	F	**7.** Trees that will remain on a lot are identified on the plot plan.
T	F	**8.** The benchmark for a lot should be located at the lowest point on the lot.
T	F	**9.** A property owner cannot build on an area where an easement has been provided.
T	F	**10.** The printreading abbreviation for dining room is DIN.

Name _____ Date _____

Printreading

Refer to the Foundation Plan on page 320.

_____ **1.** What is the length of the West foundation wall?

_____ **2.** How many risers are on the stairs leading from the basement to the main floor?

_____ **3.** What size joists are spaced 16″ OC in the basement?

_____ **4.** How many areas in the foundation are filled and tamped?

_____ **5.** How long is the steel beam for this house?

_____ **6.** How long is the total run for the stairway to the basement?

_____ **7.** How many ceiling light fixtures are located in the basement?

_____ **8.** How many 3′-0″ diameter galvanized steel areaways are required for the basement windows?

_____ **9.** The water heater is located closest to which foundation wall?

_____ **10.** What is the thickness of the vapor barrier under the main basement floor concrete slab?

_____ **11.** The drain is located closest to which foundation wall?

_____ **12.** What size concrete footings are called for beneath the pipe columns?

_____ **13.** What is the distance from the stairway to the West wall?

_____ **14.** What is the scale for the foundation plan?

_____ **15.** What size bars are used to reinforce the foundation?

_____ **16.** What is the length of the North foundation wall?

_____ **17.** The terrace contains how many square feet?

_____ **18.** What is the width of the wide flange beam?

_____ **19.** What is the center-to-center distance between the areaways on the South wall?

_____ **20.** The symbol ϕ shown on the foundation plan represents what type of light fixture?

True-False

T   F   **1.** Footings, walls, and piers are the basic features of a foundation.

T   F   **2.** The open space around a basement window is known as backfill.

T   F   **3.** Foundations are designed to support their weight only.

T   F   **4.** Information pertaining to columns and beams may be found on the foundation plan.

T   F   **5.** Foundation footings are visible on the foundation plan.

T   F   **6.** Masons and cement workers construct forms for foundations according to information given on foundation plans.

T   F   **7.** Dashed lines on each side of a foundation wall indicate the foundation footings.

T   F   **8.** Size and spacing of joists is given on the foundation plan.

T   F   **9.** A T-foundation is rarely used for residential construction.

T   F   **10.** A crawl space is normally a minimum of 30″ above ground level.

T   F   **11.** Basement stairways are not required for houses with a crawl space.

T   F   **12.** Foundation construction is the first stage in constructing a building.

T   F   **13.** When pipe columns are shown on a foundation plan, they are indicated by a dimension extending from the outside building wall to the outside wall of the column.

T   F   **14.** The direction in which floor joists run is shown on the foundation plan.

T   F   **15.** The printreading abbreviation for dishwasher is DW.

Printreading Symbols

Identify the building material shown.

_____

Identification—Foundation Plan

_____ **1.** Stairway

_____ **2.** Window and areaway

_____ **3.** Rear stoop

_____ **4.** Foundation wall

_____ **5.** Pipe column

_____ **6.** Column footing

_____ **7.** Foundation footing

_____ **8.** Steel beam

_____ **9.** Garage area

_____ **10.** Front porch

_____ **11.** Terrace

_____ **12.** Floor joists

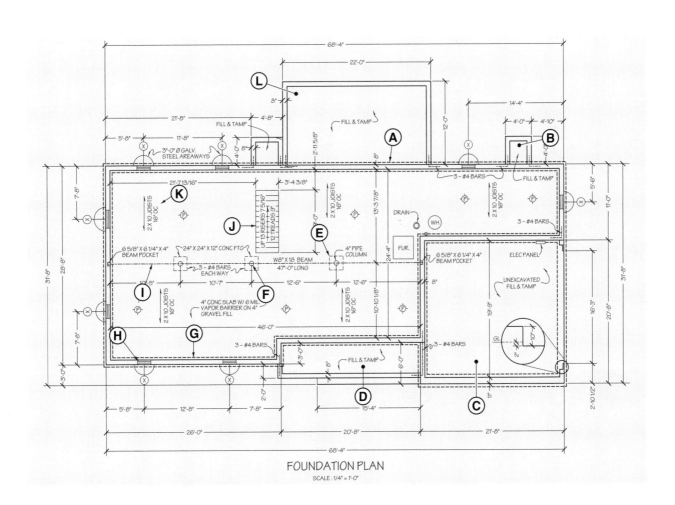

FOUNDATION PLAN

SCALE : 1/4" = 1'-0"

Name _____ Date _____

True-False

T	F	**1.** The foundation of a building must be completed before floor and wall construction begins.
T	F	**2.** Door and window openings are shown on a floor plan.
T	F	**3.** The position of interior walls is not shown on a floor plan.
T	F	**4.** Separate floor plans are usually required for each floor of a building.
T	F	**5.** Window schedules on floor plans may be designated by a circled letter.
T	F	**6.** Plumbing fixture locations are not shown on a floor plan.
T	F	**7.** The location of openings for wall heaters is given on a floor plan.
T	F	**8.** A hidden line on a floor plan may indicate a wall opening with no door.
T	F	**9.** A double wall plug is known as an electrical duplex receptacle.
T	F	**10.** Length of exterior walls on a floor plan is shown by leader lines.
T	F	**11.** Floor plans contain lines showing switch control of lights.
T	F	**12.** Electrical receptacles are not noted on a floor plan.
T	F	**13.** The floor plan of a building provides a view parallel to the floor level above the foundation.
T	F	**14.** Floor plans do not show interior wall finish details.
T	F	**15.** Pocket doors slide into a wall.
T	F	**16.** Details for kitchen cabinets are given on floor plans.
T	F	**17.** The position of exterior walls is shown on floor plans.
T	F	**18.** Information pertaining to heating methods used in a building may be found on a floor plan.
T	F	**19.** The location of hose bibbs is given on a floor plan.
T	F	**20.** The location for kitchen cabinets is given on a floor plan.
T	F	**21.** Overhead (ceiling) lights may be indicated by a symbol on a floor plan.

T F **22.** Floor plans never show access into an attic.

T F **23.** Arrows are used on a floor plan to show the direction in which ceiling joists run.

T F **24.** The size and type of roofing material for a building is given on a floor plan.

T F **25.** The printreading abbreviation for a door is DO.

Printreading

Refer to the Floor Plan on page 321.

_____ **1.** Access to the attic is located in the ceiling of the ___.

_____ **2.** Flagstone covering the terrace is ___″ thick.

_____ **3.** The floor plan is drawn to a scale of ___.

_____ **4.** The dimensions of the closet in the bedroom nearest the living room are ___.

T F **5.** Each of the three bedrooms has one closet.

_____ **6.** How many duplex receptacles are located in the living room?

_____ **7.** What is the unit rise for each step of the stairway?

_____ **8.** Overall bedroom dimensions are given from the ___.
    A. outside of exterior walls to the inside of interior walls
    B. inside of exterior walls to the inside of interior walls
    C. outside of exterior walls to the center of interior walls
    D. center of exterior walls to the center of interior walls

_____ **9.** A(n) ___ light is shown in the kitchen ceiling.

_____ **10.** The concrete slab on the terrace is ___″ thick.
    A. 2
    B. 3
    C. 5
    D. none of the above

_____ **11.** What is the size of the access to the attic?

_____ **12.** The center-to-center dimension of the garage windows is ___.

_____ **13.** The living room can be illuminated by a ___.
    A. ceiling fixture controlled by two three-way switches
    B. ceiling fixture controlled by two single switches
    C. lamp plugged into a wall receptacle controlled by two three-way switches
    D. lamp plugged into a wall receptacle controlled by a single switch

_____ **14.** How many exterior hose bibbs are shown?

_____ **15.** The bathroom entered from a bedroom has a ___.
        A. tub
        B. shower
        C. shower-tub combination
        D. none of the above

_____ **16.** What is the size of the living room?

_____ **17.** The thickness of the concrete slab on the front porch is ___″.

_____ **18.** The garage floor slopes ___ in 20′-8″.

_____ **19.** Can the family room be entered directly from the terrace?

_____ **20.** What is the length of the plant box?

T   F   **21.** Three commodes are shown on the main floor.

_____ **22.** What room can be entered from the garage?

_____ **23.** The joists in the living room ___.
        A. are spaced 16″ OC
        B. run north and south
        C. are 2 × 6s
        D. all of the above

_____ **24.** Can the kitchen be entered from the outside without going through another room?

_____ **25.** What size joists are called for in the garage?

_____ **26.** The smallest bedroom in the house measures ___ × ___.

_____ **27.** Windows in the kitchen ___.
        A. are located on the North wall
        B open inward
        C. overlook the driveway
        D. none of the above

T   F   **28.** The garage is entered from the East.

_____ **29.** What is the width of the stoop located near the NE corner of the house?

_____ **30.** How many ceiling lights are shown in the family room?

T   F   **31.** Doors entering the bedrooms from the hallway open into the bedrooms.

_____ **32.** How many wall receptacles are located in the hall walls?

_____ **33.** The stairway has ___ treads.

_____ **34.** What is the center-to-center distance of the two S windows in the family room?

_____ **35.** The overall length of the house is ___.

_____ **36.** How many walls in the living room have no windows, doors, or cased openings?

_____ **37.** What is the center-to-center distance of the R windows in the West wall?

_____ **38.** The garage has ___ electric wall receptacle(s).

_____ **39.** Partition-to-partition dimensions are shown from ___ to ___ of partitions.
            A. inside; inside
            B. outside; outside
            C. inside; center
            D. center; center

_____ **40.** What is the tread width shown on the rear stoops?

_____ **41.** Refrigerator space is provided ___.
            A. to the left of the sink
            B. on the North wall
            C. across from the oven space
            D. all of the above

_____ **42.** The garage has parking space for ___ cars.

T   F   **43.** Dining space is provided in the kitchen.

T   F   **44.** The family room is entered through double doors from the kitchen.

_____ **45.** How many exterior hinged doors are located on the North wall of the house?

Name _____ Date _____

True-False

T   F   **1.** An exterior elevation provides a view from the top of an object.

T   F   **2.** Downspouts and roof gutters are shown on elevation drawings.

T   F   **3.** The North wall of a house is considered the front wall for printreading purposes.

T   F   **4.** Elevation drawings specify the finish for interior walls.

T   F   **5.** Foundation footings on elevations are shown with hidden lines.

T   F   **6.** Elevations give actual appearance of all exterior doors and windows.

T   F   **7.** Elevation drawings include the wall surfaces and roof of a building.

T   F   **8.** Elevation drawings are always referred to by the direction in which a wall faces.

T   F   **9.** Material finish for outside surfaces of walls and roofs is given on elevation drawings.

T   F   **10.** Splashblock location for downspouts is given on elevation drawings.

T   F   **11.** Diagonal bracing required for exterior walls is indicated by solid lines on elevation drawings.

T   F   **12.** The printreading abbreviation for furnace is FURN.

Printreading

Refer to the Exterior Elevations on pages 322–323.

_____   **1.** What size bevel siding is used to cover the East elevation?

_____   **2.** How many cutting plane lines are shown in the East elevation?

_____   **3.** A(n) ___ extends through the main roof.

_____   **4.** The ground level elevation near the downspout is ___.

_____   **5.** The ground level elevation at the step leading to the South porch is ___.

T   F   **6.** Alternated vertical siding above the garage is 1″ × 8″ and 1″ × 10″.

T   F   **7.** The foundation footing is shown below the frost line.

_____   **8.** What finish material is used to face the plant box?

_____   **9.** How many shutters are shown on the South elevation?

_____   **10.** The garage roof has a rise of ___″ per foot.

T   F   **11.** The driveway is shown in the East elevation.

_____   **12.** How many rear doors are shown?

T   F   **13.** The South elevation shows one downspout.

_____   **14.** What scale is shown for the South elevation plan?

_____   **15.** ___ shingles are used to cover the roof.

_____   **16.** The rise per foot for the main roof is ___″.

T   F   **17.** Three areaways are shown in the South elevation.

T   F   **18.** Patio doors are shown in the South elevation.

_____   **19.** How many windows are shown in the East elevation?

_____   **20.** What is the size of the wood shutters required for the garage?

_____   **21.** How many lights are shown in the South entrance door?

_____   **22.** What type of roof is shown for the garage?

_____   **23.** How many window shutters are shown on the East elevation?

_____   **24.** What is the total height between the subfloor and ceiling joists of the main floor?

T   F   **25.** All elevation drawings for this house are drawn to the same scale.

_____   **26.** How many cutting plane lines are shown in the South elevation?

_____   **27.** How many downspouts are shown on the South elevation?

_____   **28.** What letter designation is given to basement windows?

_____   **29.** Bevel siding is shown applied in a(n) ___ position.

_____   **30.** The attic space of this house is ventilated by ___.

_____   **31.** What section is taken on the East elevation?

_____   **32.** What is the elevation of the terrace?

_____   **33.** What size barge rafter is called for on the West elevation?

_____   **34.** How many areaways are shown on the East elevation?

_____   **35.** What is the thickness of the concrete floor in the basement?

Name _____ Date _____

True-False

T F **1.** Section view drawings must relate to other parts of the print.

T F **2.** The arrows terminating cutting plane lines point away from the cut made.

T F **3.** The roof slope is omitted from all section views.

T F **4.** Section views are used to show interior parts of a wall.

T F **5.** A transverse section cut is made across a building.

T F **6.** Cutting plane lines are often identified by numerals.

T F **7.** Wire mesh in a slab is shown by broken lines.

T F **8.** A sill plate is also known as a secondary sill.

T F **9.** A section view of a foundation wall shows the shape of the foundation wall.

T F **10.** Grade line is often designated by the abbreviation "Gr Ln" on a print.

T F **11.** Header joists are nailed into the ends of regular joists.

T F **12.** A section view of a wall shows the height, thickness, and shape of the wall.

T F **13.** A longitudinal section cut is made along the length of a building.

T F **14.** The spacing and number of rebars in column footings may be given on a section view.

T F **15.** Cross bridging is placed between joists near the ends of their spans.

T F **16.** The size and spacing of ceiling joists are given on a print.

T F **17.** Roof overhang is the vertical distance from the side of the house to the end of the rafters.

T F **18.** Section views may be taken on floor plans and foundation plans.

T F **19.** The size of trim for the cornice may be shown on a section view.

T F **20.** The printreading abbreviation for a double hung window is DHW.

Printreading

Refer to Section A-A, Section B-B, and Section C-C on pages 324–325.

_____ **1.** The concrete floor in the basement is placed over a layer of ___″ crushed stone.

_____ **2.** Bracing in the attic with 2 × 4 purlins attached is spaced ___ OC.

T   F   **3.** Insulation is shown in the studded exterior walls.

_____ **4.** What size concrete footing is shown beneath the pipe columns?

T   F   **5.** Drain tile is placed above the concrete footings around the foundation walls.

_____ **6.** What is the OC horizontal spacing of #4 bars in the foundation walls?

_____ **7.** What size fascia board is shown?

T   F   **8.** Doubled top plates are shown on exterior walls.

_____ **9.** The double hung window sash height from the subfloor is ___.
    A. 4′-8″
    B. 6′-8½″
    C. 7′-2⁷⁄₁₆″
    D. 8′-1″

_____ **10.** Roofing materials for this house are applied in what order?
    A. building paper, sheathing, asphalt shingles
    B. sheathing, building paper, shakes
    C. sheathing, shakes
    D. sheathing, building paper, asphalt shingles

T   F   **11.** Three #4 bars run continuously through the foundation footings.

_____ **12.** What size reinforced mesh is required for the concrete floor?

_____ **13.** Collar beams in the attic are spaced ___″ OC.

_____ **14.** Are the cornices boxed in on the North wall of this house?

_____ **15.** What is the thickness of the plywood soffit?

T   F   **16.** Waterproofing is applied to both sides of all foundation walls.

_____ **17.** Every ___ joist is extended to tie in the rafters.

_____ **18.** The sill plate beneath the main floor is ___.
    A. doubled
    B. a 2 × 4
    C. a 2 × 6
    D. a 2 × 8

T   F   **19.** Cement plaster is applied to the outside of the plant box.

_____**20.** What size collar beams are used in the attic?

T   F   **21.** Flooring paper is placed over the subfloor and covered by the finish flooring.

_____**22.** What type of joint is shown between the foundation footings and walls?

T   F   **23.** Base and shoe molding are applied on partitions.

_____**24.** What material is used as backfill around drain tiles?

_____**25.** Thru-house Section A-A is drawn at a scale of ___.
- A. $3'' = 1'\text{-}0''$
- B. $\frac{1}{4}'' = 1'\text{-}0''$
- C. $\frac{1}{8}'' = 1'\text{-}0''$
- D. none of the above

T   F   **26.** An air space is provided between the sheathing and masonry veneer as shown in Section C-C.

_____**27.** Common brick is shown in Section A-A ___.
- A. on the North foundation wall
- B. on the West foundation wall
- C. above the plant box
- D. below the plant box

_____**28.** What is the rise per foot of run for the roof?

_____**29.** The cornice nailing blocks are ___.
- A. staggered
- B. offset
- C. 2×4
- D. $12''$ long

_____**30.** What is the size of the steel trap door shown in Section A-A?
- A. $\frac{1}{8}'' \times 1'' \times 24''$
- B. $\frac{1}{8}'' \times 24'' \times 24''$
- C. $1'' \times 24'' \times 24''$
- D. no steel trap door is shown

_____**31.** Roof rafters are spaced ___$''$ OC.

T   F   **32.** Pipe columns that are $4''$ in diameter support 2×10 joists.

T   F   **33.** Wall studs are spaced $24''$ OC.

_____**34.** What size anchor bolts are shown extending through the sill plate?

_____**35.** The approximate ground level elevation shown near the plant box is ___$'$.

T   F   **36.** Insulation is placed between joists in the attic.

_____**37.** What is the size of the W.F. beam shown?

_____ **38.** Reinforcing ___ is shown in the concrete slab of the front porch.

_____ **39.** Bars are spaced ___″ OC vertically in the foundation walls.

_____ **40.** What size concrete footing is specified in Garage Section B-B?

_____ **41.** Garage Section B-B is drawn at the scale of ___.

_____ **42.** Studs for the garage wall are spaced ___″ OC.

_____ **43.** How many #4 bars are shown in the foundation footing of the garage?

_____ **44.** The garage floor consists of a(n) ___″ layer of concrete placed over ___″ of crushed stone.

_____ **45.** Ceiling joists in the garage are ___s spaced ___″ OC.

Printreading Symbols

Identify this symbol.

_____

Name _____ Date _____

True-False

T	F	**1.** The distance between the countertop and wall cabinets is a standard dimension and is not given on a set of prints.
T	F	**2.** Pocket sliding doors slide into a space framed in the walls.
T	F	**3.** Countertops for base cabinets are normally 36″ above the floor.
T	F	**4.** Detail drawings may show the size and shape of molding for interior finish of a house.
T	F	**5.** Detail drawings must be drawn for all types of windows in a particular building.
T	F	**6.** Framing plans are often included as part of a set of prints.
T	F	**7.** Details show smaller objects more clearly than plan drawings.
T	F	**8.** The standard depth for wall cabinets is 14″.
T	F	**9.** Examples of construction features that may be shown on details include door and window units, fireplaces, and roof cornices.
T	F	**10.** Information pertaining to windows may be found on floor plan and elevation drawings.
T	F	**11.** The printreading abbreviation for downspout is DS.

Printreading—Kitchen Cabinet Elevations

Refer to the Kitchen Cabinet Elevations on page 325.

_____ **1.** How many base cabinet drawers are shown on the cabinet details?

_____ **2.** What is the width of the refrigerator space?

_____ **3.** The dishwasher shown is ___.
   A. 24″ wide
   B. located to the right of the sink
   C. 27″ wide
   D. a dishwasher is not shown

_____ 4. Refrigerator wall cabinets are ___″ high.

_____ 5. What is the depth of the base cabinets, excluding doors?

_____ 6. The base cabinets shown are modular in ___″ increments.
     A. 3
     B. 6
     C. 9
     D. 12

_____ 7. What is the total length of base cabinets along the North wall?

_____ 8. Finish floor-to-ceiling height in the kitchen is ___.

_____ 9. The oven cabinet on the South wall is ___″ wide.

_____ 10. This kitchen has ___ drawers 24″ wide.

_____ 11. Wall cabinets over the cooktop are ___″ high.

_____ 12. The base cabinet containing the sink is ___″ wide.

_____ 13. The countertop is ___″ deep.

_____ 14. How many lineal feet of soffit are required for the cabinets?

    T   F   15. Shelves are shown in all base cabinets.

Printreading—Framing Plans

Refer to the Framing Plans on pages 326–327.

_____ 1. The East elevation framing plan is drawn at a scale of ___.

_____ 2. What size W.F. beam is shown running lengthwise through the house?

_____ 3. The header above the garage door consists of two ___.
     A. 2 × 8s
     B. 2 × 10s
     C. 2 × 12s
     D. 2 × 14s

_____ 4. What size ridge board is shown for the main roof?

_____ 5. What size ridge board is shown for the garage?

_____ 6. What scale is shown for the front elevation framing plan?

    T   F   7. Five window openings are shown in the front elevation.

_____ 8. Every third ceiling joist is extended to serve as a(n) ___ tie.

_____ 9. Roof rafters are spaced ___″ OC.

_____ **10.** Valley rafters are shown in the ___ plan.
          A. North elevation framing
          B. ceiling joist framing
          C. roof rafter framing
          D. all of the above

_____ **11.** Louver space is shown in the ___ elevation framing plan(s).
          A. South
          B. North
          C. West
          D. all of the above

_____ **12.** What is the size of the stairway opening?

    T   F **13.** Bearing partitions are not shown in the ceiling joists framing plan.

_____ **14.** What size valley rafters are shown?

_____ **15.** Double joists are placed under all ___ running parallel to the floor joists.

Printreading—Stairway Details

Refer to the Stairway Details on page 328.

_____ **1.** This stairway has ___ of headroom.

_____ **2.** How many stringers are required?

_____ **3.** The stairwell opening is ___ in length.

_____ **4.** How many risers are shown on this stairway?

Printreading Symbols

Identify the building material shown.

_____

Name _____ Date _____

True-False

T   F   **1.** A full set of working drawings usually includes a door schedule and a window schedule.

T   F   **2.** A door schedule always contains a code letter by which doors are identified.

T   F   **3.** The locations of all door and window openings are shown on the foundation plan.

T   F   **4.** Information such as the location of a door and operating instructions is often given in the "Remarks" portion of a door schedule.

T   F   **5.** A 1″ clearance is usually allowed at the sides and top of a doorjamb.

T   F   **6.** The sizes of doors and windows are always listed on the floor plan next to the openings.

T   F   **7.** All door and window schedules list rough opening sizes.

T   F   **8.** Window schedules give similar types of information for windows as door schedules give for doors.

T   F   **9.** Molding at the bottom of walls is known as floor molding.

T   F   **10.** Door and window schedules are standard and consequently do not vary with different building plans.

T   F   **11.** The number of lights in a window is equal to the number of panes of glass plus one.

T   F   **12.** The dimensions of an entire window unit after the glass has been set in its frame is equal to the rough opening dimension.

T   F   **13.** Trim refers to molding around window and door openings.

T   F   **14.** A door schedule gives the size of doors required but not the quantity, as this may be determined from the floor plans.

T   F   **15.** The printreading symbol for drain is DR.

Printreading

Refer to the Floor Plan on page 321 and the Door and Window Schedules on page 329.

_____ **1.** A(n) ___ hollow core door is shown between the utility room and the garage.

T   F   **2.** All R windows are located in the bedrooms.

_____ **3.** What type of door is the L door?

_____ **4.** How many H doors are required?

_____ **5.** Areaway windows are set in a(n) ___ opening.

_____ **6.** What size rough opening is required for the patio doors?

_____ **7.** Recessed doors are indicated on the floor plan with the letter ___.

_____ **8.** How many flush hollow core interior doors are required?

_____ **9.** The ___ door is an E type door.

_____ **10.** What is the size of the rear service doors?

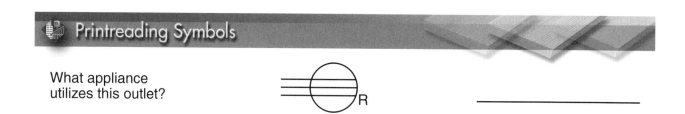

Printreading Symbols

What appliance utilizes this outlet?

_____

Unit 32

Building Codes, Zoning, Permits, and Inspections

Name _____ Date _____

Math

1. A small manufacturing company is looking for land in order to build a new facility containing 34,000 sq ft. A 270′ × 300′ lot is located. Zoning in this area will not allow the building to occupy over 40% of the land area. Is this lot large enough for the building?

2. Residential zoning regulations in a suburban community require that not more than 37.5% of the lot be occupied by a one-family dwelling. What is the maximum size one-story house that can be built on a 7400 sq ft lot?

3. A residential lot is 165′ deep. A 30′ × 64′ house is set back 44′ from the street. What is the depth of the backyard?

4. A rectangular lot for a one-family dwelling contains 14,400 sq ft. The lot is 96′ wide. How deep is the lot?

5. A contractor purchases a 150l × 280l lot for a new office for the contracting firm. The lot costs $1.50/sq ft. What is the total cost of the lot?

Short Answer

1. Building codes establish minimum standards for what purposes?

_____

_____

_____

_____

2. What are the three major types of zones in larger communities?

_____

_____

_____

3. Discuss the responsibilities of planning commissions.

_____

_____

_____

4. Who must apply to local building authorities for a building permit?

_____

_____

5. List the sequence of inspection for a one-family dwelling required by the Uniform Building Code.

_____

_____

_____

True-False

T   F   **1.** Building codes establish maximum building standards that can be used.

T   F   **2.** Model codes serve as models that can be adopted by states or local communities.

T   F   **3.** Minimum property standards establish minimum requirements for buildings constructed under the Department of Housing and Urban Development (HUD) housing programs.

T   F   **4.** Zoning regulations are laws or regulations that restrict the types of buildings that can be constructed in different areas.

T   F   **5.** Zoning regulations never change after they are established.

T   F   **6.** A building inspector has no enforcement authority.

T   F   **7.** A foundation inspection takes place after the concrete is placed in the forms.

T   F   **8.** A frame inspection takes place after the floors, walls, ceilings, and roof are framed and all blocking and bracing is installed.

T   F   **9.** In addition to a building permit, separate electrical and plumbing permits are usually required.

T   F   **10.** The printreading abbreviation for drywall is DRY.

Name _____ Date _____

Multiple Choice

_____ **1.** Builder's levels are available with telescopes ranging from ___ to ___ power.
A. 6; 16
B. 12; 16
C. 12; 32
D. 6; 32

_____ **2.** The tripod of a builder's level may have ___ legs.
A. adjustable
B. extension
C. three
D. all of the above

_____ **3.** Transit-levels have a ___ or clamp to hold the telescope in a fixed position.
A. lock-lever
B. tangent screw
C. leveling screw
D. all of the above

_____ **4.** The engineer's leveling rod is graduated in ___.
A. feet
B. feet and tenths of a foot
C. feet, tenths of a foot, and hundredths of a foot
D. none of the above

_____ **5.** The telescope of a transit-level can be ___.
A. tilted vertically
B. tilted horizontally
C. rotated vertically
D. all of the above

_____ **6.** The vernier scale of a transit-level has ___ graduations of 0 to 60 minutes at each side of the zero index.
A. 3
B. 4
C. 6
D. 12

_____ **7.** ___ are used to express fractions of a degree.
      A. Hours, minutes, and seconds
      B. Hours and minutes
      C. Minutes and seconds
      D. Minutes

_____ **8.** The horizontal circle of a transit-level ___.
      A. is turned by hand
      B. does not move when the telescope rotates
      C. is divided into four quadrants
      D. all of the above

_____ **9.** Transit-levels have a vertical ___ for measuring vertical angles.
      A. bar
      B. arc
      C. reference
      D. none of the above

_____ **10.** The telescope of a builder's level is normally adjusted with ___ leveling screws while checking the spirit level.
      A. two
      B. four
      C. six
      D. eight

Completion

_____ **1.** A tool commonly used to check and establish grades and elevations and to set up level points is a(n) ___.

_____ **2.** A ___ is used when a builder's level must be set up over a specific point.

_____ **3.** The builder's level is mounted on a three-leg support known as a ___.

_____ **4.** The builder's level has ___ and ___ crosshairs inside the barrel of the telescope.

_____ **5.** A horizontal ___ screw allows the telescope of a builder's level to be moved slightly to the left or right.

_____ **6.** Leveling rods are made of wood, plastic, or ___.

_____ **7.** A horizontal ___ screw is used to hold the builder's level in a fixed horizontal position.

_____ **8.** A(n) ___ is the vertical measuring device held by a second person when a builder's level is used to check or establish grades and elevations.

_____ **9.** A(n) ___ leveling rod is graduated in feet, inches, and eighths of an inch.

_____ **10.** Red ___ numbers are the largest numbers on a leveling rod.

_____ **11.** All elevation points on a job site relate to a benchmark or ___ that is established for the job.

_____ **12.** The figures and graduations between foot numbers on a leveling rod are generally printed in the color ___.

_____ **13.** The intersecting scales of a transit-level used to measure horizontal angles are the ___ and ___ scales.

_____ **14.** The angle most often used in construction work is the ___ angle.

_____ **15.** ___ leveling screws are used to make a rough adjustment for an automatic level.

_____ **16.** The printreading abbreviation for East is ___.

Identification—Arm Signals

_____ **1.** Lower target

_____ **2.** Move rod to right

_____ **3.** Target is on grade

_____ **4.** Rod is plumb

_____ **5.** Raise target

_____ **6.** Move rod to left

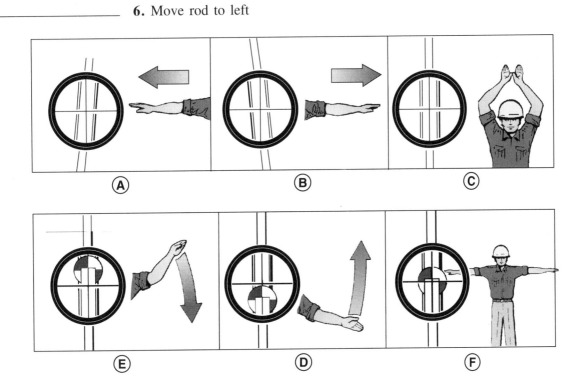

Short Answer

1. What precautions should be taken with all leveling instruments?

_____

_____

_____

_____

_____

_____

_____

2. List step-by-step procedures to follow when leveling a builder's level.

_____

_____

_____

_____

_____

_____

_____

3. Discuss the general procedure for finding the difference between two grade points.

_____

_____

_____

_____

_____

_____

4. List the steps for laying out a 90° angle with a transit-level.

_____

_____

_____

_____

_____

_____

_____

_____

Identification—Transit-Level

_____ **1.** Focusing knob

_____ **2.** Leveling vial

_____ **3.** Eyepiece

_____ **4.** Leveling screw

_____ **5.** Plumb bob hook

_____ **6.** Horizontal graduated circle

_____ **7.** Horizontal tangent screw

Laser Levels and Total Station Instruments

Name _____ Date _____

Math

1. 36°-12′ + 15°-32′ =

2. 120° + 17°-58′ =

3. 90° + 15°-7′-35″ =

4. 60°-3′-38″ + 15°-12′ =

5. 180°-19′-16″ + 12°-30′-24″ =

6. 145° − 32′-38″ =

7. 96°-52′ − 35′-20″ =

8. 270° − 90°-45′ =

9. 58′-40″ − 16′-31″ =

10. 45° − 38°-35′ =

11. 17°-32′ + 91°-15′-30″ =

12. 112°-20′-35″ + 9°-15′-35″ =

13. 150° – 45°-44′ =

17. 45′-50″ – 30′-20″ =

14. 88°-28′ – 60°-15′-45″ =

18. 130°-15′ – 45°-40′ =

15. 36°-37′-45″ + 52′ =

19. 120°-40′-30″ – 90°-55′ =

16. 245° + 15° + 38°-12′ =

20. 180° + 70° + 55°-15′ =

True-False

    T   F   **1.** The laser level requires only one person to perform any layout operation.

    T   F   **2.** Strong air disturbances can affect the accuracy of a laser light.

    T   F   **3.** The beam of a laser level is approximately 1⅜″ in diameter.

    T   F   **4.** A laser level can be used to plumb horizontal items.

    T   F   **5.** A laser level is accurate to within ¹⁄₁₆″ at a range of 100′.

    T   F   **6.** Laser levels may be wall-mounted for certain operations.

    T   F   **7.** A laser level is used for establishing grades and leveling over long distances.

    T   F   **8.** A manual-leveling laser level is leveled in a manner similar to traditional surveying equipment.

    T   F   **9.** The target for a laser level is known as a detector.

    T   F   **10.** A compensated self-leveling laser level has fully automatic leveling operations.

    T   F   **11.** Class II lasers that are commonly used on construction sites can sometimes present a health hazard.

    T   F   **12.** Total station instruments combine survey technology with digital data processing.

T   F   **13.** Electronic distance measurement has an accuracy of 0.1′ without the use of a measuring tape.

T   F   **14.** A reflectorless total station can measure distance more accurately than a total station that uses a reflector.

T   F   **15.** The printreading abbreviation for elevation is ELV.

Identification—Laser Level

_____   **1.** Beacon cover

_____   **2.** Decreases beacon speed

_____   **3.** Base

_____   **4.** Out-of-level indicator

_____   **5.** Rotating beacon

_____   **6.** Leveling bubble

_____   **7.** Increases beacon speed

_____   **8.** Battery life indicator

_____   **9.** Battery housing

_____   **10.** Leveling screw

_____   **11.** ON/OFF switch

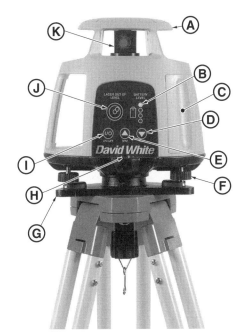

David White Instruments

Building Site and Foundation Layout

Name _____ Date _____

Multiple Choice

_____ **1.** Information needed to lay out foundation walls is found in the ___.
    A. floor plan
    B. plot plan
    C. elevations
    D. specifications

_____ **2.** In the building trades, the term *frost line* refers to the ___.
    A. northernmost point at which soil normally freezes
    B. southernmost point at which soil normally freezes
    C. depth to which soil freezes
    D. none of the above

_____ **3.** Building site features that help determine the type of foundation include ___.
    A. shape, size, and slope of the lot
    B. weather conditions
    C. soil conditions
    D. all of the above

_____ **4.** The distances from the property lines to the building are known as ___.
    A. setbacks
    B. recesses
    C. offsets
    D. spacings

_____ **5.** The depth to which trenches for foundation footings must be dug is usually found in the ___ of the foundation plan.
    A. floor plan
    B. plot plan
    C. finish schedule
    D. section views

_____ **6.** Batter boards are normally placed ___' to ___' behind corners to provide working room for form construction.
    A. 2; 4
    B. 4; 6
    C. 6; 8
    D. 8; 10

_____ **7.** A ___ line on a plot plan shows the shape of the varying grades of the lot.
  A. ground
  B. soil
  C. profile
  D. contour

_____ **8.** A paved area around the foundation of a building should slope ___″ per ___′.
  A. ⅛; 1
  B. ¼; 1
  C. ½; 1
  D. 1; 1

_____ **9.** ___ are constructed to hold the building lines during foundation form work.
  A. Contour lines
  B. Batter boards
  C. Lookouts
  D. Hubs

_____ **10.** The recommended minimum slope for unpaved areas around a building is ___″ in ___′.
  A. 4; 8
  B. 4; 10
  C. 6; 8
  D. 6; 10

Completion

_____ **1.** A piece of property in a residential area with established streets is known as a(n) ___.

_____ **2.** Silt particles are larger than ___ particles.

_____ **3.** Settlement can be expected to occur in any newly constructed building unless it is built on ___.

_____ **4.** Reinforcing steel bars are normally required in concrete or masonry foundation walls constructed in ___ risk zones.

_____ **5.** Most buildings are constructed on soils that are classified as ___, ___, ___, or ___.

_____ **6.** Building lots must be ___ to determine precise boundaries before foundation construction begins.

_____ **7.** Foundation ___ must be placed below the frost line to prevent movement of the foundation during freezing and thawing of the soil.

_____ **8.** Sand particles compress more than ___ particles when subjected to heavy pressure.

_____ **9.** Foundations must be designed to withstand greater ___ in areas where earthquakes might occur.

_____ **10.** Local ___ normally specify the depth of foundation footings.

_____ **11.** The property point, or ___ point, is identified by a small nail driven into the corner stake.

_____ **12.** A shallow ___ should be cut in a batterboard to prevent the lines from moving.

_____ **13.** A wooden stake or a pipe with a lead plug may be used for ___ stakes.

_____ **14.** The printreading abbreviation for excavate is ___.

Name _____ Date _____

Identification—Foundations

_____ d _____	**1.** Rectangular foundation
_____ G _____	**2.** Battered pier
_____ B _____	**3.** Battered foundation
_____ C _____	**4.** L-shaped foundation
_____ A _____	**5.** T-shaped foundation
_____ E _____	**6.** Stepped foundation
_____ F _____	**7.** Wood post
_____ H _____	**8.** Wood post block

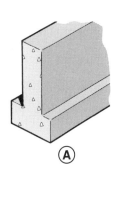

(A)

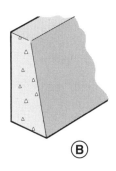

(B)

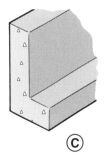

(C)

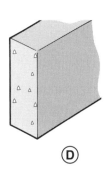

(D)

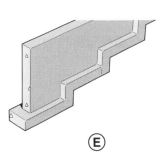

(E)

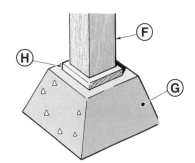

(F) (H) (G)

True-False

T   F   **1.** A footing is the base for a wall.

T   F   **2.** Rectangular foundations are designed for buildings with heavy wall loads and loose soil conditions.

T   F   **3.** Pressure-treated lumber or redwood is recommended for use as mudsills because of its decay and insect resistance.

T   F   **4.** A common code requirement is that the bent end of an anchor bolt be placed at least 7″ into unreinforced masonry.

T   F   **5.** Sills for interior framed walls may be attached to the floor slab with a powder-actuated fastener.

T   F   **6.** The bottoms of floor joists in a crawl-space foundation are normally 18″ or more above the ground.

T   F   **7.** Vertical footings of stepped foundations must be at least 8″ thick.

T   F   **8.** The Uniform Building Code requires that anchor bolts be within 12″ of the ends of any piece of sill plate.

T   F   **9.** Footings spread the weight of a building over a wider area.

T   F   **10.** A slab-at-grade foundation features a framed floor unit.

T   F   **11.** Concrete piers are square, round, or hexagonal in shape.

T   F   **12.** Foundation sills are normally constructed of 2 × 4s or 2 × 6s.

T   F   **13.** Anchor bolts are also known as J-bolts.

T   F   **14.** Grade beams are foundation walls that receive their main support from stepped footings.

T   F   **15.** Anchor bolts used to fasten sill plates should be at least ¾″ in diameter.

T   F   **16.** Anchor bolts used to fasten sills to concrete slabs must be set in the concrete at the time of the pour.

T   F   **17.** Washers should always be used under nuts on anchor bolts.

T   F   **18.** Areaways must project below the finish grade and above the bottom of the window.

T   F   **19.** Stepped foundations cannot be used with a full basement.

T   F   **20.** The printreading abbreviation for exterior is EXT.

Completion

Sill plate   **1.** ___ are normally attached to the top of a foundation wall to provide a nailing area for joists or studs directly on the foundation.

shims ?   **2.** ___ may be applied on foundation walls to provide an even, level base for sills.

_____   **3.** The bottoms of girders supporting floor joists in a crawl-space foundation are normally ___″ or more above the ground.

twice   **4.** The gravel bed for a wood foundation must be at least ___ as wide as the footing plate.

_____   **5.** The ___, ___, and ___ are the three main types of foundation systems.

_____   **6.** The moisture content of plywood used in wood foundations should not exceed ___%.

_____   **7.** In a full-basement foundation, the basement wall should extend at least ___″ above the finish grade.

_____   **8.** The moisture content of framing lumber used in wood foundations should not exceed ___ %.

stepped   **9.** A ___ foundation is normally used on a steeply sloped lot.

_____   **10.** Foundation ___ are combined with a concrete floor slab in slab-at-grade foundations.

_____   **11.** Basement walls in a full-basement foundation are normally ___′ to ___′ high.

Name _____ Date _____

Math

_____ **1.** ___ cubic yards equals 297 cu ft.

_____ **2.** ___ of concrete are required to pour a 21' × 36' concrete slab 6" thick.

_____ **3.** A #12 rebar is ___" larger in diameter than a #8 rebar.

_____ **4.** A concrete wall that is 8' high, 24' long, and 8" thick requires ___ yd of concrete.

_____ 12 _____ **5.** A form with an area of 27 sq ft can be poured ___" deep with 1 cu yd of concrete.

_____ **6.** In a concrete mix that contains 1½ cu ft of gravel, ½ cu ft of water, 1 cu ft of cement, and 2½ cu ft of sand, ___% of the mix is sand.

_____ 1 _____ **7.** ___ yd of concrete is/are needed to pour a 6" slab containing 54 sq ft.

_____ **8.** At $76.35/cu yd, what is the concrete cost for a slab containing 7 yd of concrete?

_____ **9.** ___ full yards of concrete selling for $72/yd can be purchased for $1800.

_____ **10.** A #4 rebar is ___" smaller in diameter than a #6 rebar.

True-False

T ⓕ **1.** Proportions of cement, aggregate, and water for concrete are consistent, regardless of the mix required.

T   F **2.** Print specifications may give the types of admixtures for concrete.

T   F **3.** Concrete should normally be discharged from a concrete truck within 1½ hr after water has been added to the mixture.

T   F **4.** When pouring foundation walls, concrete should be poured in lifts of 12" to 20".

T   F **5.** Sand is the fine aggregate used in concrete mixtures.

T   F **6.** The process of manufacturing cement was developed by Joseph Portland.

T   F **7.** Concrete has more compression strength than lateral strength.

T   F **8.** Floor slabs are more difficult to cure than walls.

T	F	**9.** Rebar may be placed in the cavities or mortar beds of concrete-block walls to help prevent cracking.
T	F	**10.** Wire mesh is identified by its weight per running foot.
T	F	**11.** A #8 rebar is 1″ in diameter.
T	F	**12.** Rebar size and placement are shown on the plot plan.
T	F	**13.** Wire mesh is placed in forms for concrete slabs, walks, and driveways to help prevent cracking of the concrete.
T	F	**14.** Forms for concrete walls should remain in place for three to seven days to allow sufficient curing time.
(T)	F	**15.** Concrete containing rebar is known as reinforced concrete.
T	F	**16.** The printreading abbreviation for face brick is FCB.

Completion

_____ 1. The strongest and most durable material used for foundations is ___.

_____ gravel _____ 2. The largest part of a concrete mixture consists of fine and coarse ___.

_____ 3. Gravel or crushed stone used in a concrete mixture ranges from ___″ to ___″ in diameter.

_____ 4. Concrete walls that are narrow or have closely spaced rebar require a concrete mixture that has ___ aggregate.

_____ 5. Vertical and ___ pressure are the different kinds of pressure exerted on a foundation wall.

_____ 6. The ___ strength of concrete is largely determined by its water-cement ratio.

_____ 7. Transit-mix concrete trucks have drum capacities ranging from ___ cu yd to ___ cu yd.

_____ 8. One cubic yard contains ___ cu ft.

_____ 9. ___ concrete is delivered to the job site by truck.

_____ 10. Consolidating freshly poured concrete helps eliminate open spaces known as ___.

_____ 11. The placing of concrete is known as a(n) ___.

_____ 12. ___ concrete trucks have the capability of mixing concrete while en route to the job site.

_____ 13. Mechanical ___ are used to consolidate concrete on larger jobs.

_____ 14. ___ is the process of keeping concrete moist to allow proper hydration.

_____ 15. ___ is the hardening reaction that occurs between water and cement in a concrete mix.

Name _____ Date _____

Identification—Job-Built Forms

_____ **1.** Bottom plate

_____ **2.** Gravel

_____ **3.** Stake

_____ **4.** Temporary brace

_____ **5.** Keyway

_____ **6.** Drain tile

_____ **7.** Footing

_____ **8.** Double walers

_____ **9.** Diagonal brace

_____ **10.** Studs

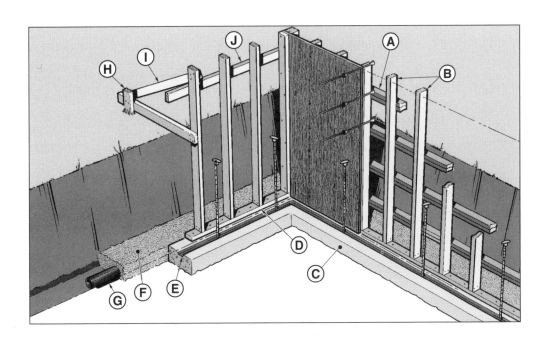

True-False

T	F	**1.** Concrete forms are temporary structures.
T	F	**2.** Spread footings for T-shaped walls are generally formed with 4″ thick planks.
T	F	**3.** Plyform® is available in thicknesses ranging from ¼″ to 1¾″.
T	F	**4.** Particleboard is the most commonly used material for sheathing form walls.
T	F	**5.** Slow pours of concrete into wall forms increase strain on the forms more quickly than fast pours.
T	F	**6.** Duplex nails should be used on forms whenever practical.
T	F	**7.** Stakes are normally spread 4′ apart when ¾″ plywood is used as sheathing for low walls.
T	F	**8.** The pressure from poured concrete increases as the height or thickness of a wall form decreases.
T	F	**9.** The distance between walers in a wall form is determined by the thickness and height of the wall to be poured.
T	F	**10.** Frames constructed on the job for door and window openings in concrete walls are known as bucks.
T	F	**11.** Panel form snap ties are laid out 2′ OC.
T	F	**12.** Small wall forms may be tied together with wood cleats or braces.
T	F	**13.** Door and window bucks are removed after the concrete has set up and before the frames are stripped.
T	F	**14.** Snap ties can be snapped off at breakback points after the forms are stripped.
T	F	**15.** Panel form sections are always constructed on the job site.
T	F	**16.** Walers, studs, and braces are normally cut from 2 × 6s.
T	F	**17.** Spreader cones used with snap ties and steel wedges maintain the distance between walls.
T	F	**18.** Panel forms are generally considered more efficient than built-in-place forms.
T	F	**19.** When 2″ thick planks are used for sheathing, walers are not required.
T	F	**20.** The printreading abbreviation for finish is FNSH.

Name _____   Date _____

Identification—Insulating Concrete Forms

_____   **1.** Waffle grid

_____   **2.** Block

_____   **3.** Plank

_____   **4.** Panel

_____   **5.** Flat core

_____   **6.** Screen grid

(A)   (B)   (C)

(D)   (E)   (F)

Multiple Choice

_____   **1.** The concrete footings and walls of a low T-foundation are poured ___.
   A. footings first
   B. walls first
   C. at the same time
   D. in no particular order

_____   **2.** The footings and walls of a T-foundation having low walls and a crawl space
   are poured ___.
   A. as one unit
   B. footings first
   C. walls first
   D. in no particular order

_____   **3.** Rectangular or battered forms can be built by the ___ method.
   A. built-in-place or panel
   B. monolithic
   C. trench
   D. all of the above

_____ **4.** The printreading abbreviation for finish floor is ___.
    A. FNSH FL
    B. FIN F
    C. FNS F
    D. none of the above

_____ **5.** Key strips used to form keyways in footings are placed in forms ___.
    A. before the pour and are not removed after the concrete hardens
    B. before the pour and are removed after the concrete hardens
    C. during the pour and are not removed after the concrete hardens
    D. during the pour and are removed after the concrete hardens

_____ **6.** The bottoms of piers should always ___.
    A. be larger than the tops
    B. be the same size as the tops
    C. rest on firm soil
    D. none of the above

_____ **7.** Forms for T-foundation footings may not be required where ___.
    A. temperatures do not fall below 32°F for more than 10 consecutive days
    B. the structure has a southern orientation
    C. the soil condition is firm and stable
    D. none of the above

_____ **8.** Keyways in T-foundations help secure ___.
    A. batterboards to sheathing
    B. stakes to panels
    C. walls to footings
    D. none of the above

_____ **9.** Insulating concrete form wall designs include ___.
    A. flat core
    B. waffle grid
    C. screen grid
    D. all of the above

_____ **10.** Rebars for tall T-foundation walls are placed ___.
    A. after the outside form walls are set
    B. before the outside form walls are set
    C. after the inside form walls are set
    D. none of the above

_____ **11.** ___ are square, round, or battered structures that serve as a base for wood posts or steel columns.
    A. Steel pipe columns
    B. Piers
    C. Cleats
    D. none of the above

_____ **12.** Wall forms for T-foundations are built ___.

    A. when forms for the footings are built

    B. before the concrete has set up in the footings

    C. after the concrete has set up in the footings

    D. none of the above

_____ **13.** The sections of a patented panel system are secured to each other with ___.

    A. bolts and nuts

    B. wedge bolts or clamps

    C. wood screws

    D. duplex nails

_____ **14.** Piers that serve as a base for steel columns may be ___.

    A. round

    B. square

    C. battered

    D. all of the above

_____ **15.** Common types of insulating concrete forms are ___.

    A. block

    B. panel

    C. plank

    D. all of the above

Matching

_____ **1.** Insulating concrete forms

_____ **2.** Monolithic

_____ **3.** Keyway

_____ **4.** Pour strip

_____ **5.** Pier

_____ **6.** Post-tensioned slabs

_____ **7.** Precast grade beam

_____ **8.** Key strips

_____ **9.** Rebars

_____ **10.** Pillowing

A. poured at the same time

B. piece of wood to indicate proper level of concrete pour

C. chamfered 2 × 4s used to form keyways

D. concrete between foam layers

E. base for wood post or steel column

F. reinforcing steel bars

G. concrete members formed at a plant

H. groove in a footing

I. high-strength tendons

J. bulges in wall

Name _____ Date _____

Completion

_____ **1.** Dimensions for patios are found on the ___ plan of blueprints.

_____ **2.** Concrete ___ for sidewalks, patios, and driveways are usually placed directly on the ground.

_____ **3.** Forms for outdoor slabs are normally constructed of ___ or ___.

_____ **4.** In most areas, ___ or cement masons set forms for outdoor slabs.

_____ **5.** After concrete has been placed and struck off, ___ perform finishing operations on concrete slabs.

_____ **6.** A(n) ___ board, or strike board, may be used to strike off a concrete slab.

_____ **7.** ___ joints provide spacing between dissimilar construction.

_____ **8.** Driveways may be reinforced with ___ or ___ to help prevent cracking.

_____ **9.** ___ are grooves cut into concrete slabs to control cracking.

_____ **10.** Double-car driveways are normally ___′ to ___′ wide.

True-False

T   F   **1.** The minimum recommended thickness for sidewalks is 6″.

T   F   **2.** Screeding operations may be performed manually.

T   F   **3.** Patios should have a minimum slope of 1″ in 12′ to provide drainage.

T   F   **4.** Local building codes normally specify minimum widths for sidewalks.

T   F   **5.** Concrete may rot under exceedingly damp conditions.

T   F   **6.** Spreader boards may be left in place after concrete is poured.

T   F   **7.** Forms for walks and driveways should not be set up until final soil grading has been completed.

T   F   **8.** Driveways should have a minimum cross slope of ½″ per foot for drainage.

T   F   **9.** Driveways should be at least 4″ thick.

T   F   **10.** The printreading abbreviation for fireplace is FP.

Unit 41

Foundation Moisture Control and Insect Prevention

Name _____ Date _____

True-False

T	F	**1.** Water tables normally rise during wet seasons and drop during dry seasons.
T	F	**2.** Water rises higher in porous soils than in less porous soils.
T	F	**3.** Vapor barrier films should be covered with pea gravel, sand, or concrete.
T	F	**4.** Basement foundation walls should be waterproofed from the edge of the footing to the finish-grade line.
T	F	**5.** All types of drain pipes and drain tiles should be surrounded by a layer of gravel.
T	F	**6.** Sump wells may be drained by gravity or a pump may be utilized to remove water in the well.
T	F	**7.** All plastic drain pipe is perforated to allow seepage.
T	F	**8.** Vapor barriers must be decay-resistant.
T	F	**9.** The highest point below ground that is normally saturated with water is known as the water table.
T	F	**10.** Vapor barriers are normally constructed of 6 mil polyethylene.
T	F	**11.** The process by which water rises is known as sepilary action.
T	F	**12.** Subterranean termites account for 95% of all termite damage in the United States.
T	F	**13.** Nonsubterranean termites live above the ground.
T	F	**14.** Moisture control is an effective method for preventing termite damage.
T	F	**15.** Termite shields should be bent upwards at a 45° angle.
T	F	**16.** Termites consume the interior portions of wooden boards.
T	F	**17.** Metal termite shields are required by all local building codes in the United States.
T	F	**18.** Vapor barriers must be resistant to insect attack.
T	F	**19.** Subterranean termites live in underground nests.
T	F	**20.** Termite shields should extend at least 4″ on each side of a foundation wall.
T	F	**21.** Galvanized iron is used for termite shields.

T   F   **22.** In some areas, chemicals may be applied directly to the soil to help prevent termites.

T   F   **23.** Chemicals in pressure-treated lumber increase resistance to termites.

T   F   **24.** Asphalt sheet membranes are sheets of rubberized asphalt laminated to the foundation wall.

T   F   **25.** The printreading abbreviation for fireproof is FPRF.

Floor Framing

Name _____ Date _____

Matching

_____ 1. Built-up beam

_____ 2. Floor joists

_____ 3. Blocking

_____ 4. Pocket

_____ 5. Allowable span

_____ 6. Load-bearing beam

_____ 7. Lally

_____ 8. Underpinning

_____ 9. Post

_____ 10. Cantilevered

A. projecting beyond a support

B. steel pipe column

C. recess in foundation wall

D. two or more planks

E. supports weight of load and wall above

F. vertical wood or metal support

G. framework on which subfloor rests

H. members that strengthen and maintain joist spacing

I. clear span of a beam

J. short studded walls in platform construction

Completion

_____ 1. Joists should be doubled under ___ that run in the same direction as the joists.

_____ 2. The ___ is the wood deck that rests on the floor joists.

_____ 3. A wood post should be equal to the width of the ___ it supports.

_____ 4. The two major factors affecting the size of joist material are ___ and ___.

_____ 5. To maintain 16″ OC floor joists, the first joist is marked ___″ from the edge of the building.

_____ 6. Top and bottom members of a wood floor truss are known as ___.

_____ **7.** ___ may be used to extend the height of low foundation walls.

_____ **8.** A(n) ___ is a recess in a foundation wall that receives a beam.

_____ **9.** ___ joists are located at each side of a floor opening.

_____ **10.** The chords of a typical floor truss are tied together by ___ members.

_____ **11.** Subfloor panels may have square edges or ___ edges.

_____ **12.** Subfloor panels up to ⅞″ thick are generally nailed with ___d common nails.

Multiple Choice

_____ **1.** The minimum clearance between the bottom of floor joists and the ground is ___″.
  A. 12
  B. 18
  C. 24
  D. 36

_____ **2.** Joints between the planks of a built-up beam are ___.
  A. staggered
  B. adjacent
  C. mitered
  D. none of the above

_____ **3.** Bridging is normally required when joist spans exceed ___′.
  A. 6
  B. 8
  C. 10
  D. 12

_____ **4.** Beams for post-and-beam subfloor systems are generally placed ___″ OC.
  A. 16
  B. 24
  C. 32
  D. 48

_____ **5.** Floor trusses are normally placed ___″ OC.
  A. 12
  B. 18
  C. 24
  D. 48

_____ **6.** Header joists are also known as ___ joists.
     A. inner or outer
     B. interior or exterior
     C. rim or band
     D. horizontal or level

_____ **7.** Post caps may be attached to wooden posts and beams by ___.
     A. nails and bolts
     B. adhesive and mastic
     C. welding and fitting
     D. all of the above

_____ **8.** Load-bearing beams must support the ___.
     A. wall framed directly above
     B. live load of the floor system directly above
     C. dead load of the floor system directly above
     D. all of the above

_____ **9.** Underpinning is used for ___.
     A. crawl-space foundations only
     B. extending the height of a low foundation wall
     C. steel beams only
     D. all of the above

_____ **10.** Web stiffeners are ___ used to reinforce an I-joist web.
     A. dimension lumber
     B. oriented strand board (OSB)
     C. structural panel materials
     D. all of the above

_____ **11.** ___ joists are used when a floor or balcony extends past the wall below it.
     A. Run-on
     B. Extension
     C. Projection
     D. Cantilevered

_____ **12.** A subfloor is also known as ___ flooring.
     A. base
     B. finish
     C. rough
     D. crown

_____ **13.** Minimum spacing between subfloor panels is ___″ for end joints and ___″ for edge joints.
     A. $\frac{1}{16}$; $\frac{1}{16}$
     B. $\frac{1}{16}$; $\frac{1}{8}$
     C. $\frac{1}{8}$; $\frac{1}{8}$
     D. $\frac{1}{8}$; $\frac{1}{16}$

Short Answer

1. What is the purpose of underlayment?

_____

_____

2. List three methods for attaching floor underlayment.

_____

_____

3. List five methods for using short joists for long spans.

_____

_____

_____

_____

_____

_____

_____

4. What are the advantages of a floor truss system?

_____

_____

_____

_____

5. What is the primary advantage of using solid blocking instead of header joists to support joists over exterior walls?

_____

_____

_____

Printreading Symbols

Identify this control device.

（T）_____

Identification—Wood Beam on Post

_____ **1.** Pocket

_____ **2.** Post

_____ **3.** Foundation wall

_____ **4.** Sill plate

_____ **5.** Built-up wood beam

_____ **6.** Angle iron bracket

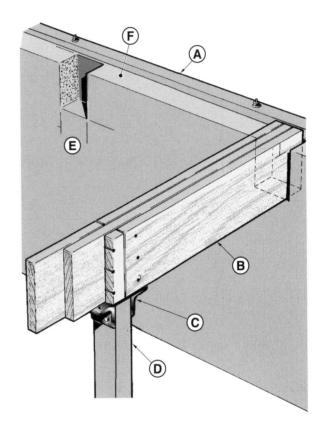

True-False

T F **1.** Non-load-bearing beams must support the dead load and the live load of the floor system directly above.

T F **2.** The minimum depth of a pocket in a foundation wall is 4½″.

T F **3.** The crown in a floor joist should be turned down.

T F **4.** Plywood is the oldest type of panel product used for subfloors.

T F **5.** The printreading abbreviation for fixture is FXTR.

T F **6.** Floor underlayment is located beneath the subfloor.

T F **7.** Subfloor panels are laid with the long edge parallel to the floor joists.

T	F	**8.** Floor trusses are generally prefabricated.
T	F	**9.** The bow shape in a joist is known as crown.
T	F	**10.** Wood cross bridging is normally attached to floor joists by toenailing.
T	F	**11.** Tail joists run from a header to a supporting wall or beam.
T	F	**12.** Floor joists are normally placed 24″ OC.
T	F	**13.** A post-and-beam subfloor receives its main support from floor joists.
T	F	**14.** Floor trusses are normally spaced 24″ OC.
T	F	**15.** Steel pipe columns are also known as lallys.

Math

_____ **1.** A 41′ steel post is centered between two foundation walls. A beam with an allowable span of 12′-0″ is set into pockets in the foundation wall. What is the length of the beam?

_____ **2.** A house foundation requires 142′ of 2″ × 4″ sill plate. At $.46 per lineal foot, what is the cost of the sill plate?

_____ **3.** How wide must a pocket be to provide ½″ air space on each side of a built-up beam of three 2 × 12s?

_____ **4.** How many sheets of ¾″ × 4′ × 8′ plywood are required for a 24′ × 32′ subfloor?

_____ **5.** Twenty-eight sheets of ¾″ × 4′ × 8′ plywood are required to deck a subfloor. At $18.87 per sheet, what is the cost of the plywood?

_____ **6.** An 8′-1½″ mudsill has ___ studs placed 16″ OC.

_____ **7.** Thirty-two 2″ × 4″ × 3′-0″ studs are required for extending the height of a low foundation wall. How many 12′ long 2 × 4s are required to cut the 3′-0″ studs?

_____ **8.** A 24′ × 28′ subfloor can be laid at a cost of $2.86/sq ft. What is the total material cost for the subfloor?

Name _____ Date _____

Identification—Exterior Stud Wall

_____ **1.** Outside corner post

_____ **2.** Bottom plate

_____ **3.** Window header

_____ **4.** Cripple stud

_____ **5.** Rough sill

_____ **6.** Inside corner

_____ **7.** Door header

_____ **8.** Structural panel

_____ **9.** Double top plate

_____ **10.** Trimmer stud

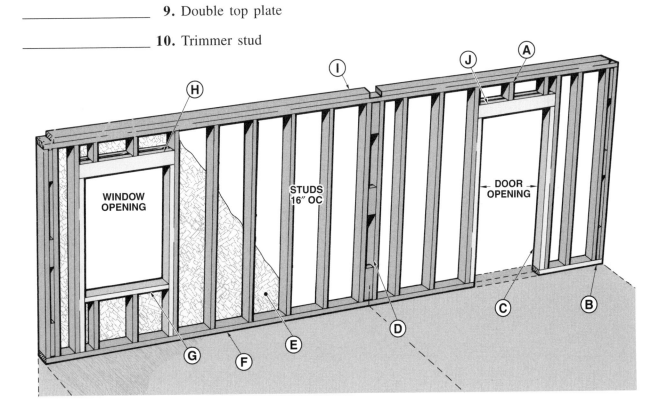

WINDOW OPENING

STUDS 16" OC

DOOR OPENING

True-False

T F **1.** Non-load-bearing partitions run in the same direction as the joists.

T F **2.** The most commonly used structural timber for one-story buildings is a 2 × 6.

T F **3.** Exterior walls of a house divide the living area into separate rooms.

T F **4.** Load-bearing partitions support the ends of floor joists or ceiling joists.

T F **5.** Corner posts are constructed wherever one wall ties into another wall.

T F **6.** Wall studs are normally spaced 16″ OC.

T F **7.** Outside corners occur at the ends of a wall.

T F **8.** A doubler strengthens the upper section of a wall and helps carry the weight of floor and ceiling joists.

T F **9.** Inside corners occur at the ends of a wall.

T F **10.** The standard height of walls in wood-framed houses is 8′ from subfloor to ceiling joists.

T F **11.** Window and door headers may be built-up or one solid piece.

T F **12.** Cripple studs are normally placed 24″ OC.

T F **13.** Headers should be placed at the top of rough openings for doors or windows.

T F **14.** A nominal 2″ × 4″ stud wall is 3½″ wide.

T F **15.** The standard height of doors is 6′-8″.

T F **16.** The tops of all door and window openings in all walls of a single-story building are generally in line with one another.

T F **17.** Cripple studs for headers are normally required only where walls are over 8′-1″ high.

T F **18.** Diagonal bracing of let-in type utilizes 1″ × 6″ boards for strength.

T F **19.** Diagonal bracing should be placed at the ends of walls and at 8′ intervals.

T F **20.** Diagonal bracing is normally let-in the walls after the wall section is raised.

T F **21.** Exterior walls that are properly covered with structural sheathing do not require diagonal bracing.

T F **22.** Fire blocks should be nailed at the midpoint of a wall.

T F **23.** Fire blocks may be nailed in a straight line or staggered.

T F **24.** The first step in laying out walls by the horizontal plate layout method is cutting wall plates.

T   F   **25.** To ensure alignment of standard size panels on exterior walls, the first stud from the corner of a wall should be placed 15¼″ from the end of the corner.

T   F   **26.** Prints may give rough or finish dimensions for door openings.

T   F   **27.** The rough height to the top of a door is the vertical distance from the subfloor to the top of the door header.

T   F   **28.** The printreading abbreviation for floor is FLR.

T   F   **29.** When raising wall sections, short sections should be raised first.

T   F   **30.** Wall sections may be plumbed before raising if accurate measurements are made.

T   F   **31.** The tops of wall sections must be aligned before the corners are plumbed.

T   F   **32.** Oriented strand board (OSB) may be used for sheathing exterior walls.

T   F   **33.** Metal fasteners may be used with wood-framed walls.

T   F   **34.** Plywood sheathing may be added to a squared wall before or after the wall is raised.

T   F   **35.** Plywood sheathing may be placed with the grain running horizontally or vertically.

Completion

_____ **1.** Wall construction of a house begins after the ___ has been fastened in place.

_____ **2.** ___ are the vertical framing members that run between wall plates.

_____ **3.** The horizontal framing members directly above a stud wall comprise the double ___.

_____ **4.** The plate at the bottom of a wall is the ___ plate, or the bottom plate.

_____ **5.** Interior walls are also known as ___.

_____ **6.** Headers are supported by ___ studs, which fit between the sole plate and the bottom of the header.

_____ **7.** ___ studs between the header and double top plate of a door opening help carry the weight from the top plate to the header.

_____ **8.** The distance between a rough ___ and frame is determined by the window, frame, and vertical clearance dimensions required.

_____ **9.** Diagonal bracing provides ___ strength for a wall.

_____ **10.** Cripple studs are required if headers are less than ___″ wide.

_____ **11.** ___ blocks are nailed into a wall to slow down a fire traveling inside the wall.

_____ **12.** A(n) ___ pole may be used in the vertical layout of a wood-framed wall.

_____ **13.** All ends of wall sections should be fastened with ___d nails.

_____ **14.** The bottom plate of a wall section should be nailed through the wood subfloor into the floor ___ to provide rigidity.

_____ **15.** Exterior wall ___ is the material used for exterior covering of outside walls.

Printreading Symbols

Identify the electrical symbol shown.

(F)

_____

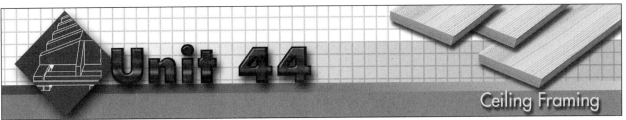

Unit 44

Ceiling Framing

Name _____ Date _____

True-False

T F **1.** The size of a ceiling joist is determined by the pitch of the roof.

T F **2.** The ceiling frame ties together the outside walls of a building.

T F **3.** Butted ceiling joists must be cleated with plywood or a metal strap.

T F **4.** A low-pitched hip roof requires stub joists in the hip section.

T F **5.** In a two-story building, the ceiling of the lower story serves as the floor unit for the upper story.

T F **6.** Ceiling joists should run in the same direction as roof rafters whenever possible.

T F **7.** Ribband ends are nailed to inside walls and partitions of a building.

T F **8.** When spacing is the same between ceiling joists and studs, the ceiling joists should be offset halfway between the studs.

T F **9.** When ceiling joists are perpendicular to the roof rafters, their outside ends must be cut to the slope of the roof.

T F **10.** When roof rafters are spaced 24″ OC and joists are spaced 16″ OC, every other rafter will be placed next to a ceiling joist.

T F **11.** Joist ends trimmed for the roof slope should be cut on the crown side.

T F **12.** A flat roof should have a pitch of ¼″ per foot or more to shed water.

T F **13.** A scuttle over 3′ square must have double joists and headers.

T F **14.** Lookout rafters of a flat roof are cantilevered.

T F **15.** The printreading abbreviation for opening is OPNG.

Completion

_____ **1.** If a beam is placed above the ceiling joists, the ___ end of the joists are hung from the beam.

_____ **2.** Spacing for ceiling joists is normally ___″ or ___″ OC.

_____ **3.** Ceiling joists that overlap an interior bearing partition should overlap at least ___″.

_____ **4.** One end of a ceiling joist must rest on a(n) ___ wall.

_____ **5.** The use of a(n) ___ is the most effective method to prevent twisting or bowing of ceiling joists.

_____ **6.** A flat roof is required to withstand a(n) ___ of up to 40 lb/sq ft.

_____ **7.** ___ are the most important framing members of the ceiling.

_____ **8.** A(n) ___ or ___ is used to prevent twisting or bowing of ceiling joists.

_____ **9.** Walls that run in the same direction as ceiling joists require ___, or deadwood.

_____ **10.** An opening framed in the ceiling to provide access to the attic is known as a(n) ___.

Printreading Symbols

Identify the symbol shown.

SD

_____

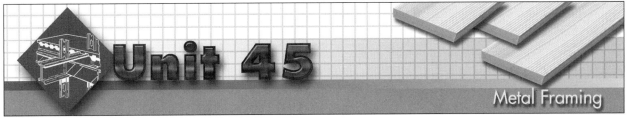
Name _____ Date _____

Completion

_____ **1.** Light-gauge framing receives a protective hot-dip ___ treatment.

_____ **2.** Self-tapping screws include self-drilling and ___ screws.

_____ **3.** The printreading abbreviation for overhead door is ___.

_____ **4.** Sheathing materials are attached to a metal-framed wall with ___.

_____ **5.** The most popular metal stud widths are ___″ and ___″.

_____ **6.** Metal joist sizes range from ___ to ___.

_____ **7.** Metal-framed roof trusses may be assembled on the job site or they may be ___ in a shop.

_____ **8.** After welding galvanized steel components, a(n) ___ coating must be applied to the weld area.

_____ **9.** ___ is the most commonly used interior wall finish for metal-framed walls.

_____ **10.** The thicknesses of load-bearing metal-framing wall members range from ___ ga to ___ ga.

Short Answer

1. Explain how a subfloor is attached to metal floor joists.

_____

_____

_____

_____

_____

2. List five advantages of metal framing materials.

_____

_____

_____

_____

3. List two common methods of joining metal structural members.

_____

_____

_____

_____

_____

Identification—Floor Unit Components

_____ **1.** Floor to foundation connection

_____ **2.** Cantilevered joists

_____ **3.** Joists lapped over beam

_____ **4.** Cross bridging

_____ **5.** Solid blocking

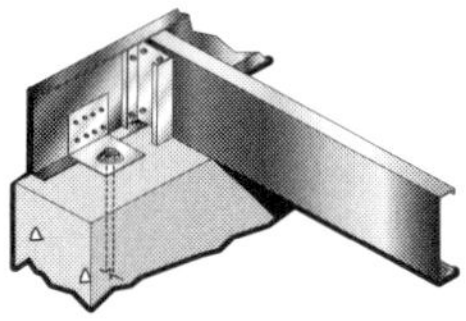

(A)

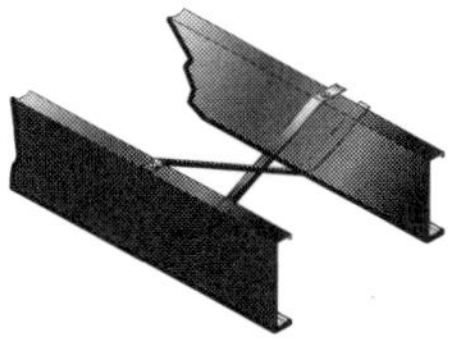

(B)

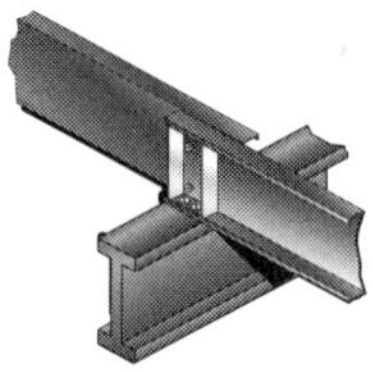

(C)

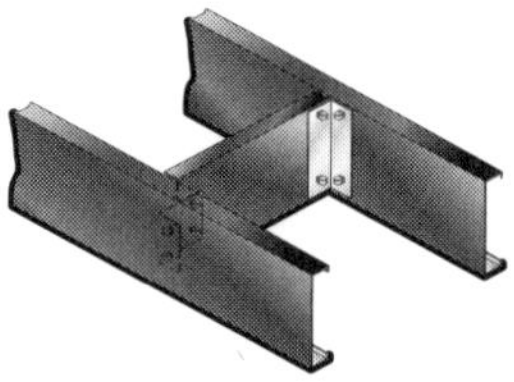

(D)

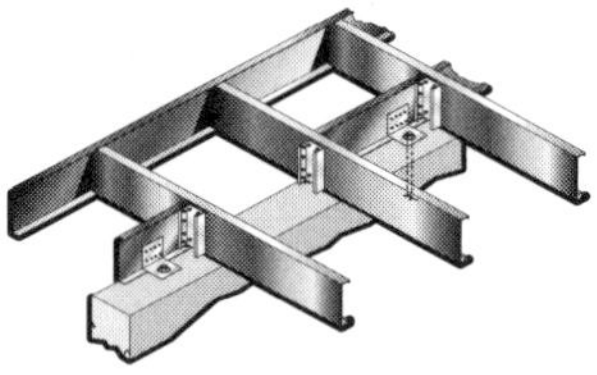

(E)

Identification—Steel Shapes

_____ **1.** Track

_____ **2.** U-channel

_____ **3.** C-shape

_____ **4.** Furring channel

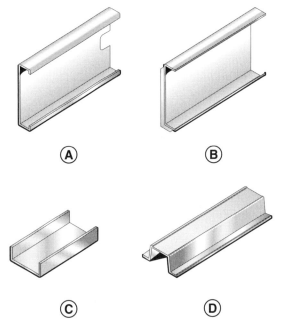

Identification—Screw Head Types

_____ **1.** Hex washer

_____ **2.** Truss

_____ **3.** Oval

_____ **4.** Wafer

_____ **5.** Modified truss (lath)

_____ **6.** Bugle

_____ **7.** Pan

_____ **8.** Round washer

_____ **9.** Hex

_____ **10.** Trim

_____ **11.** Round

_____ **12.** Pancake

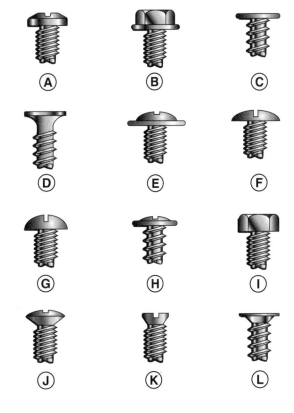

Name _____ Date _____

Matching

_____ **1.** Pitch

_____ **2.** Total span

_____ **3.** Unit run

_____ **4.** Total run

_____ **5.** Total rise

_____ **6.** Dead load

_____ **7.** Live load

_____ **8.** Allowable span

_____ **9.** Unit rise

_____ **10.** Sheathing

A. 12″

B. slope of a roof

C. actual roof height

D. weight of materials used to construct the roof

E. one-half of the total span

F. weight and pressure of wind and snow

G. base for finish roof material

H. rise per foot

I. width of a building

J. distance from ridge to outside wall plates

Math

_____ **1.** What is the total rise of a roof with a 32′ span and a 6″ unit rise?

_____ **2.** What is the total rise of a roof with a 28′ span and a 7″ unit rise?

_____ **3.** What is the total rise of a roof with a 30′ span and a 5″ unit rise?

_____ **4.** What is the total rise of a roof with a total run of 16′ and a 5″ unit rise?

_____ **5.** A roof has a total rise of 8′-6″ and a span of 34′. What is the unit rise?

_____ **6.** A roof has a unit rise of 5″ and a total run of 15′. What is the total span?

_____ **7.** A roof has a 6″ unit rise and a total span of 26′. What is the total run?

_____ **8.** A roof has an 8″ unit rise and a total run of 16′. What is the total rise?

_____ **9.** A print shows the unit rise and unit run for a 32′ total span roof as [diagram: 12″ / 4″]. What is the total rise for this roof?

_____ **10.** A print shows the unit rise and unit run for a 30′ total span roof as [diagram: 12″ / 6″]. What is the total rise for this roof?

True-False

T   F   **1.** A hip roof is easier to construct than a gable roof.

T   F   **2.** Shed roofs are primarily used for porches and additions to a building.

T   F   **3.** A mansard roof has less living space underneath than a shed roof.

T   F   **4.** A gambrel roof has more living space directly underneath than a gable roof.

T   F   **5.** Pitch refers to the span of a roof.

T   F   **6.** Proper water drainage is an important aspect of roof design.

T   F   **7.** The overall width of a building is the total run.

T   F   **8.** The printreading abbreviation for footing is FT.

T   F   **9.** The weight of sheathing and insulation may be disregarded when determining the dead load of a roof.

T   F   **10.** Flat roofs in northern states are generally required to be able to withstand live loads of 60 lb/sq ft.

T   F   **11.** Live load requirements for similar roofs may vary in different parts of the United States.

T   F   **12.** Purlins are vertical members used to strengthen long rafters.

T   F   **13.** Ceiling rafters are normally placed 16″ OC.

T   F   **14.** Flat roofs shed snow more easily than sloped roofs.

T   F   **15.** Ceiling joists hold the tops of walls in place.

T   F   **16.** Ceiling joists are normally spaced 16″ OC.

T   F   **17.** Collar ties are normally placed in the lower one-third area of the attic space.

T   F   **18.** End joints of sheathing panels should be staggered.

Completion

_____ **1.** The ___ roof is the easiest roof to build.

_____ **2.** A hip roof has ___ sloping sides.

_____ **3.** The gambrel roof has a double slope on ___ sides.

_____ **4.** Basic roof types can be combined to form ___ roofs.

_____ **5.** The total rise must be known before carpenters can set the ___ at the correct height.

_____ **6.** A(n) ___ roof slopes in one direction.

_____ **7.** The most commonly used roof for single-family dwellings is the ___ roof.

_____ **8.** Total rise, in inches, may be found by multiplying the number of feet in the total run times the unit ___.

_____ **9.** Unit rise is the number of inches that a rafter rises vertically for each ___ of unit run.

_____ **10.** A gable roof slopes in ___ directions.

_____ **11.** The ___ roof is the strongest type of roof because it is braced by four rafters.

_____ **12.** Rafters on a hip roof run at ___° angles from the corners to the ridge.

_____ **13.** The mansard roof has a double slope on ___ sides.

_____ **14.** Plywood and ___ panels are often used as roof sheathing.

_____ **15.** The recommended minimum thickness for waferboard used for roof sheathing is ___".

Identification—Roof Types

_____ **1.** Mansard

_____ **2.** Gambrel

_____ **3.** Hip

_____ **4.** Gable

_____ **5.** Shed

Multiple Choice

_____ **1.** A shed roof is also known as a ___ roof.
A. biplane
B. lean-to
C. flat
D. none of the above

_____ **2.** The ___ roof is higher on the eaves than in the center.
A. continuous-slope gable
B. mansard
C. butterfly shed
D. gambrel

_____ **3.** The ___ of a roof is the number of inches the rafters rise vertically for each foot of run.
A. unit rise
B. total rise
C. unit run
D. span

_____ **4.** Metal rafter anchors are often used to tie roof rafters to the supporting ___.
A. floor
B. wall
C. door openings
D. none of the above

_____ **5.** The hypotenuse of a right triangle is similar to the ___ of a roof.
A. total run
B. total span
C. total rise
D. common roof rafter

_____ **6.** The three basic types of roofs are the ___, ___, and ___.
A. shed; hip; gambrel
B. shed; mansard; gable
C. shed; hip; gable
D. shed; butterfly; gable

Name _____ Date _____

Completion

_____ 1. ___ rafters of a gable roof extend from the top wall plates to the ridge board.

_____ 2. A gable roof slopes in ___ directions.

_____ 3. The ___ board of a gable roof is placed at the peak of the roof.

_____ 4. The basic structural members of a gable roof are the ___ board, ___ rafters, and ___ studs.

_____ 5. ___ cuts are made at the ridge, heel, and tail of common rafters on gable roofs.

_____ 6. Rafter tables on steel squares give unit rises from ___" to ___".

_____ 7. Lengths of common rafters for gable roofs are based on the ___ rise and total ___ of the roof.

_____ 8. The unit rise and total run of a roof are found on the ___ of a building.

_____ 9. Rafter tables are printed on the ___ side of a framing square.

_____ 10. Shortening the rafters refers to deducting one-half the thickness of the ___ board from each common rafter.

_____ 11. When laying out common rafters with a steel square and a pair of square gauges, the square gauge on the tongue is secured on the number corresponding to the ___.

_____ 12. When laying out common rafters with a steel square and a pair of square gauges, the square gauge on the blade is secured on the number corresponding to the ___.

_____ 13. Plumb cuts for a common rafter must be laid out on the ___, ___, and ___ of the rafter.

_____ 14. Whenever more than one piece of material is required for a long ridge board, the break should fall at the ___ of a rafter.

_____ **15.** ___ joists for gable roofs are normally spaced 16″ OC.

_____ **16.** Rafters for gable roofs are normally spaced ___″ OC.

_____ **17.** ___ boards nailed to the tail ends of common rafters serve as a finish piece at the edge of a gable roof.

_____ **18.** A(n) ___ roof has a double slope on each side.

_____ **19.** Roofs with dormers must have a steep ___ and a high ___.

_____ **20.** A(n) ___ roof slopes in only one direction.

True-False

T   F   **1.** Level seat cuts allow rafters of a gable roof to rest on the top wall plates.

T   F   **2.** All common rafters of a gable roof are the same length.

T   F   **3.** The ridge board provides a nailing surface for the top ends of common rafters on a gable roof.

T   F   **4.** Gable studs provide a nailing surface for siding and sheathing at the end of the roof.

T   F   **5.** The run of the overhang is the vertical distance from the building line to the tail cut of a rafter for a gable roof.

T   F   **6.** Dormers for houses generally have gable or shed roofs.

T   F   **7.** The total run of a shed roof is the width of the building less one wall width.

T   F   **8.** Subflooring for attic spaces under gambrel roofs should be placed after the roof is framed.

T   F   **9.** Purlins and collar ties may be used to strengthen gable roofs.

T   F   **10.** The printreading abbreviation for foundation is FDN.

T   F   **11.** The tail plumb line on common rafters of a gable roof may be cut before or after the rafters are nailed in place.

T   F   **12.** The common length difference of gable studs may be found by using a steel framing square.

T   F   **13.** Gable studs increase in length from the ridge board to the outside walls.

T   F   **14.** Rafters and joists of gable roofs should never tie into one another.

T   F   **15.** Setting common rafters in place constitutes the major part of gable roof construction.

T   F   **16.** Ridge boards are normally wider than rafters.

T   F   **17.** The ridge board of a gable roof should be cut to length after all common rafters are nailed in place.

T   F   **18.** Common rafters for gable roofs may be precut to length before roof-framing operations begin.

T   F   **19.** The length of a ridge board equals the length of the building and the overhang at the gable ends.

T   F   **20.** The theoretical length of a common rafter is always shorter than its actual length.

Identification—Gable Dormers

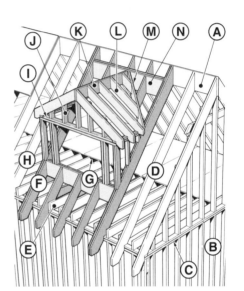

_____   **1.** Wall studs

_____   **2.** Wall double top plate

_____   **3.** Trimmer stud

_____   **4.** Floor joist

_____   **5.** Ridge board

_____   **6.** Double gable corner post

_____   **7.** Subfloor

_____   **8.** Dormer double top plate

_____   **9.** Double header

_____   **10.** Dormer gable studs

_____   **11.** Valley jack rafter

_____   **12.** Common roof rafter

_____   **13.** Valley rafter

_____   **14.** Gable dormer ridge board

Printreading Symbols

This symbol shows a(n) ____ door.

Identification—Gable Roof Framework

_____ **1.** Gable stud

_____ **2.** Common rafter

_____ **3.** Ridge board

_____ **4.** Double top plate

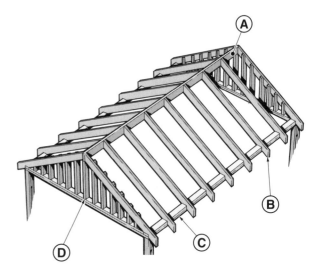

Unit 48

Hip Roofs

Name _____ Date _____

Multiple Choice

_____ **1.** Hip roofs have ___ sloping sides.

　　A. no

　　B. two

　　C. three

　　D. four

_____ **2.** Hip rafters run at a ___° angle from the corners of the building to the ridge.

　　A. 22½

　　B. 45

　　C. 60

　　D. 90

_____ **3.** The unit run for a common rafter is ___″.

　　A. 8

　　B. 10

　　C. 12

　　D. 32

_____ **4.** The *length of hip* line on a steel square rafter is also known as the ___.

　　A. side cut scale

　　B. theoretical ridge length

　　C. valley per foot run

　　D. diagonal overhang

_____ **5.** When hip roofs are framed with common rafters, one-half of the ___° thickness of the common rafter must be deducted.

　　A. 22½

　　B. 45

　　C. 60

　　D. 90

_____ 6. The run of a hip rafter overhang is ___″ for each 1″ of common rafter overhang.

      A. 1.42

      B. 14.2

      C. 42

      D. 142

_____ 7. The hip rafter overhang of a roof with a 16″ common rafter overhang is ___″.

      A. 6.72

      B. 16.72

      C. 22.72

      D. 27.72

True-False

T   F   **1.** Hip rafters of a hip roof are always longer than common rafters.

T   F   **2.** Side cuts are not required on hip rafters.

T   F   **3.** Common rafters for hip roofs are laid out in the same manner as common rafters.

T   F   **4.** Rafter tables on the steel square cannot be used to find the lengths of hip rafters.

T   F   **5.** The theoretical length of a hip rafter equals its actual length.

T   F   **6.** When laying out hip rafters, the ridge plumb and side cuts should be marked first.

T   F   **7.** The overhang of a hip rafter is longer than the overhang of a common rafter.

T   F   **8.** The side cut angle of a hip rafter varies with the unit rise.

T   F   **9.** Angles of the plumb and seat cuts on a hip rafter are the same as the angles for a common rafter.

T   F   **10.** Dropping a hip rafter is faster than backing it.

T   F   **11.** The printreading abbreviation for truss is TRS.

Completion

_____ **1.** A(n) ___ cut is required on a hip rafter where it meets the ridge board.

_____ **2.** Hip ___ rafters frame the roof between hip rafters and outside walls.

_____ 3. The unit run for a hip rafter is ___″.

_____ 4. The theoretical length of a hip rafter is the distance from the ___ plumb cut line to the center of the ridge.

_____ 5. A hip roof having a common rafter overhang of 18″ has a hip rafter overhang of ___″.

_____ 6. ___ the top edges of hip rafters may be accomplished by chamfering.

_____ 7. Hip rafters may be ___ by making the seat cut larger.

_____ 8. Hip jack rafters ___ in length as they near the end of a building.

_____ 9. The common ___ differences for hip jack rafters can be found by using the steel square rafter table.

_____ 10. Hip jack rafters require a single side cut where they fasten to the ___.

Identification—Hip Roof Framework

_____ 1. Hip jack rafter

_____ 2. Ridge board

_____ 3. Hip rafter

_____ 4. Common rafter

Name _____ Date _____

Multiple Choice

_____ **1.** The two sections of an intersecting roof ___.
A. may be the same or different widths
B. must be the same width
C. must be different widths
D. form a flat plane

_____ **2.** In a roof with equal spans, the total rise is ___.
A. different for each ridge
B. the same for each ridge
C. based upon the number of floors in the building
D. none of the above

_____ **3.** ___ rafters run from valley rafters to both ridges.
A. Valley
B. Valley jack
C. Hip-valley cripple
D. none of the above

_____ **4.** Valley rafters always run at a ___° angle to the outside walls of a building.
A. 22½
B. 45
C. 60
D. 90

_____ **5.** Valley rafters always run ___ to hip rafters.
A. parallel
B. diagonal
C. perpendicular
D. none of the above

_____ **6.** A shortened valley rafter runs at a ___° angle to the supporting valley rafter.
A. 22½
B. 45
C. 60
D. 90

_____ **7.** Valley jack rafters ___ in length as they near the top of a roof.
    A. increase
    B. decrease
    C. may increase or decrease
    D. remain constant

_____ **8.** The space between hip and valley rafters, which are placed close together, is framed with ___ rafters.
    A. common
    B. gable
    C. hip-valley cripple jack
    D. none of the above

_____ **9.** A hip-valley cripple jack rafter requires ___.
    A. a plumb cut at one end and a side cut at the other end
    B. a side cut at one end and a side cut at the other end
    C. plumb and side cuts at each end
    D. none of the above

_____ **10.** The length of a valley cripple jack rafter is ___ the valley jack rafter.
    A. one-half the length of
    B. the same length as
    C. twice the length of
    D. none of the above

Completion

_____ **1.** A(n) ___ is formed where sloping sections of an intersecting roof meet.

_____ **2.** An intersecting roof is also known as a(n) ___ roof.

_____ **3.** Valley rafters are the same length as ___ rafters.

_____ **4.** Valley rafters do not require backing or ___.

_____ **5.** The printreading abbreviation for furnace is ___.

_____ **6.** Valley jack rafters require a(n) ___ cut on the end nailed to the ridge.

_____ **7.** Valley cripple jack rafters are used only on intersecting roofs with ___ spans.

_____ **8.** ___ valley construction is a method of building intersecting roofs without valley rafters.

_____ **9.** The ridge of the smaller roof section of a roof with unequal spans is ___ than the main ridge.

_____ **10.** An intersecting roof with unequal spans requires ___ and ___ valley rafters.

Identification—Intersecting Roof

_____	**1.** Hip rafter
_____	**2.** Common rafter
_____	**3.** Main ridge board
_____	**4.** Hip jack rafter
_____	**5.** Double top plate
_____	**6.** Valley jack rafter
_____	**7.** Intersecting ridge board
_____	**8.** Valley rafter

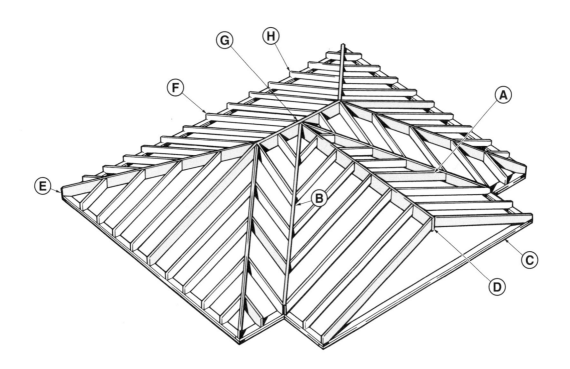

Name _____ Date _____

True-False

    T   F   **1.** The bottom chords of a roof truss serve as ceiling joists.

    T   F   **2.** Trusses require intermediate support.

    T   F   **3.** Web members of a roof truss run between the top and bottom chords.

    T   F   **4.** The printreading abbreviation for galvanized iron is GI.

    T   F   **5.** The bottom chords of a roof truss serve as roof rafters.

    T   F   **6.** Roof trusses are normally assembled on the roof.

    T   F   **7.** Wooden roof trusses have been in use only since the early part of the nineteenth century.

    T   F   **8.** A tight fit between truss members is required for the structural integrity of the truss.

    T   F   **9.** Trusses may be built on the job site.

    T   F   **10.** Trusses are not designed for buildings with hip roofs.

    T   F   **11.** Trusses may be prefabricated.

    T   F   **12.** Small, light trusses can be placed by hand on one-story buildings.

    T   F   **13.** Conventional fall protection is required when performing work at or over 6′.

    T   F   **14.** Every part of a truss is always in a state of tension or compression.

    T   F   **15.** Trussed roofs require more labor to construct than traditional framed roofs.

    T   F   **16.** Roof trusses rest on two exterior walls of a building and transfer loads to these walls.

    T   F   **17.** Local snow and wind conditions need not be considered when designing a truss.

    T   F   **18.** To prevent sagging, the bottom chord of a truss is arched when the truss is being constructed.

    T   F   **19.** Inadequate temporary bracing is the main cause of truss collapse during truss placement.

T   F   **20.** Truss height commonly ranges from 5′ to 15′.

T   F   **21.** Job-built trusses are usually fastened together with plywood connector plates.

T   F   **22.** Trusses are normally spaced 2′ OC.

Multiple Choice

_____ **1.** Over ___% of new homes in the United States are constructed with roof trusses.
   A. 35
   B. 55
   C. 75
   D. 95

_____ **2.** Residential roof trusses range from ___ ′ to ___ ′ long.
   A. 10; 26
   B. 14; 32
   C. 15; 50
   D. 24; 80

_____ **3.** The basic components of a roof truss are ___ and ___.
   A. ridges; purlins
   B. jacks; studs
   C. chords; web members
   D. all of the above

_____ **4.** ___ connectors are the most commonly used connectors for trusses in residential and light commercial construction.
   A. Metal plate
   B. Striated plastic
   C. Corrugated plyform
   D. none of the above

_____ **5.** ___ roofs are the easiest type of trussed roof to construct.
   A. Hip
   B. Gable
   C. Scissors
   D. none of the above

_____ **6.** The ___ of trusses tend to sag at the center after they have been set in place.
   A. short web members
   B. long web members
   C. top chords
   D. bottom chords

_____ **7.** The minimum size of lumber used for chords in truss fabrication is ___.

     A. 2 × 2

     B. 2 × 3

     C. 2 × 4

     D. 2 × 6

_____ **8.** The minimum size of lumber used for webs in truss fabrication is ___.

     A. 2 × 2

     B. 2 × 3

     C. 2 × 4

     D. 2 × 6

Identification—Common Truss Designs

_____ **1.** Scissors

_____ **2.** Piggyback

_____ **3.** Split

_____ **4.** Kingpost

_____ **5.** Attic frame

_____ **6.** Hip

_____ **7.** W-type (fink)

_____ **8.** Gable

Identification—Truss Components

_____ **1.** Heel

_____ **2.** Stud

_____ **3.** Peak

_____ **4.** Bottom chord

_____ **5.** Top chord

_____ **6.** Splice

_____ **7.** Web

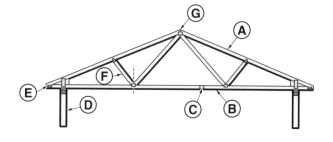

Name _____ Date _____

Multiple Choice

_____ 1. Materials with a perm rating of ___ or less qualify as vapor barriers.
      A. 0.25
      B. 1.00
      C. 2.50
      D. 5.00

_____ 2. Thermal insulation has a higher heat flow resistance than ___.
      A. brick
      B. concrete
      C. masonry
      D. all of the above

_____ 3. Insulation in buildings is normally placed by ___.
      A. sheet metal workers
      B. carpenters
      C. plumbers
      D. electricians

_____ 4. In a one-story house, insulation is normally placed in the ___.
      A. exterior walls
      B. floors
      C. ceiling
      D. all of the above

_____ 5. Vapor barriers and ventilation are needed in the ___ areas of a house.
      A. inhabited
      B. attic
      C. crawl space
      D. all of the above

_____ 6. Heat is transferred from one area to another area by ___.
      A. conduction
      B. convection
      C. radiation
      D. all of the above

_____ **7.** A forced hot air furnace moves heat by ___.
    A. conduction
    B. convection
    C. radiation
    D. none of the above

_____ **8.** ___ is not a homogenous material.
    A. Concrete
    B. A hollow concrete masonry unit
    C. Wood
    D. Polyurethane foam

_____ **9.** An underfloor radiant heating system heats a building by ___.
    A. conduction
    B. convection
    C. radiation
    D. none of the above

_____ **10.** Activities of a four-member family produce approximately ___ lb of water vapor in 24 hr.
    A. 1.12
    B. 2.25
    C. 11.1
    D. 22.5

_____ **11.** The heat from a fireplace warms an area by ___.
    A. conduction
    B. convection
    C. radiation
    D. all of the above

_____ **12.** Modern houses are generally ___ than older houses.
    A. less well insulated
    B. smaller
    C. better insulated
    D. none of the above

_____ **13.** Moisture problems due to condensation occur most often in ___ climates.
    A. arid
    B. cold
    C. warm
    D. none of the above

_____ **14.** Moisture rises from the ground into a building by ___.
    A. capillary action
    B. thermal reaction
    C. radiant induction
    D. none of the above

Completion

_____ 1. Insulation should be placed in walls, floors, and ceilings so that it surrounds the ___ of a building.

_____ 2. ___ is the movement of heat by direct transmission of invisible waves.

_____ 3. The circulation of heat in a pipe by hot water moving through it is an example of ___.

_____ 4. ___ is the process by which drops of water are formed on cool surfaces when moisture comes out of warm air.

_____ 5. Insulation materials that have vapor barriers should be placed so that the vapor barrier is attached to the ___ side of the wall.

_____ 6. ___ is the movement of heat through a solid or liquid.

_____ 7. ___ are any materials that conduct heat easily.

_____ 8. A Btu is the amount of heat required to raise the temperature of 1 lb of water ___ °F.

_____ 9. ___ is the movement of heat through circulatory motion of air or a liquid.

_____ 10. Materials that do not conduct heat easily are known as ___.

_____ 11. As air temperature increases, the moisture-holding capacity of the air ___.

_____ 12. Vapor ___ and ventilation are practical means for controlling condensation.

_____ 13. ___ vents located beneath roof cornices provide ventilation through several small openings or a continuous slot.

_____ 14. Gable end ___ are generally considered to be the oldest ventilation system used for gable roofs.

_____ 15. ___ should be placed over the inside of a louver to prevent insects from entering the building.

_____ 16. The printreading abbreviation GDR stands for ___.

Printreading Symbols

This electrical symbol shows a(n) ___ switch.

S   _____

True-False

T F **1.** Insulation and improved construction methods reduce the amount of energy required to heat or cool a building.

T F **2.** Construction materials used in the walls of a building, such as plywood panels and masonry products, aid in preventing heat flow.

T F **3.** In the summer, the outside warm air moves away from the cooler air inside a building.

T F **4.** Heated air is lighter than nonheated air.

T F **5.** Cool air rises more rapidly than heated air.

T F **6.** In the winter, heated air inside a house moves toward outside air.

T F **7.** Air expands as it is cooled.

T F **8.** Cold air moves toward hot air.

T F **9.** An effective insulator absorbs heat rather than reflects it.

T F **10.** Heat received from the sun is an example of convection.

T F **11.** Wood has a higher k factor than concrete.

T F **12.** The C factor is used when measuring conductivity of plywood panels.

T F **13.** Poor conductors make good insulators.

T F **14.** Polyurethane foam has a higher k factor than wood.

T F **15.** The total heat flow resistance of a wall is equal to the R value of insulation placed in the wall.

T F **16.** Concealed condensation problems usually cause more serious damage to a structure than surface condensation problems.

T F **17.** The k factor is not used when measuring conductance of a hollow concrete masonry unit.

T F **18.** The dew point is the temperature at which condensation occurs.

T F **19.** Metal and glass are permeable materials that do not allow vapor passage.

T F **20.** Radiant heat can warm objects in a room, which can then give off heat by radiation.

Printreading Symbols

This door swings ___.

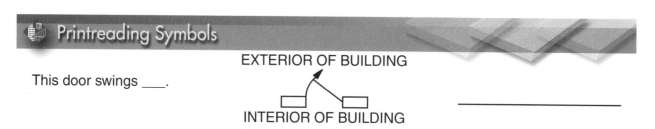

EXTERIOR OF BUILDING

INTERIOR OF BUILDING

Matching

_____ **1.** k factor

_____ **2.** C factor

_____ **3.** U factor

_____ **4.** R value

_____ **5.** Conduction

_____ **6.** Radiation

_____ **7.** Insulators

_____ **8.** Convection

_____ **9.** Dew point

_____ **10.** Conductors

A. ability of material to resist heat flow

B. measures conductivity through homogenous material

C. transmittance measurement of how many Btu/hr pass through combination materials

D. measures conductivity through non-homogenous material

E. movement of heat through circulatory motion of air or liquid

F. movement of heat through a solid or liquid

G. materials that transfer heat quickly

H. materials that do not transfer heat quickly

I. temperature at which condensation occurs

J. direct transmission of heat by invisible waves

Name _____ Date _____

Completion

_____ **1.** Thermal insulation materials resist ___ flow from a building during cold weather.

_____ **2.** ___ insulation is commercially available in bags and may be poured in place directly from the bag.

_____ **3.** Insulation is rated by its ___ value.

_____ **4.** ___ are highly effective multilayer insulating wall systems that continue to gain wider acceptance in commercial and residential construction.

_____ **5.** ___ blanket and batt insulation stays in place by friction.

_____ **6.** Rigid foam insulation is usually applied on the ___ walls of a building.

_____ **7.** ___ conducts 6 to 10 times more Btu per hour than an equivalent area of framed wall.

_____ **8.** Faced insulation often has a(n) ___ foil vapor barrier.

_____ **9.** Wood or metal ___ doors may be hung on the outside of an exterior door frame to aid in the prevention of heat loss.

_____ **10.** Heat loss through windows and doors occurs primarily as a result of ___.

_____ **11.** Foamed-in-place insulation is poured or ___ into wall cavities.

_____ **12.** Thermal insulation materials resist ___ flow into a building during warm weather.

_____ **13.** Buildings in cold climates require insulation with higher ___ values than insulation used for buildings in warmer climates.

Printreading Symbols

This light is controlled by a(n) ___ switch.

S ◯   _____

Multiple Choice

_____ 1. ___ percent of the energy used in an average home is for heating and cooling the building.
    A. Thirty
    B. Fifty
    C. Seventy
    D. Ninety

_____ 2. Doors may be sealed against heat loss by ___.
    A. caulking
    B. weather-stripping
    C. installing thresholds
    D. all of the above

_____ 3. Heat loss in a residence occurs primarily through the ___.
    A. roof
    B. walls
    C. foundation
    D. windows and doors

_____ 4. Heat loss through slab-at-grade floors can be substantially reduced by installing ___″ to ___″ of rigid foam insulation.
    A. ½; 1
    B. 2; 4
    C. 4; 6
    D. 6; 8

_____ 5. Batt lengths of flexible insulation are commercially available in widths of ___″ and ___″.
    A. 12; 16
    B. 12; 24
    C. 16; 24
    D. all of the above

_____ 6. ___ insulation is the material most widely used to insulate walls, floors, ceilings, and attics.
    A. Blanket and batt
    B. Rigid foam
    C. Loose fill
    D. Foamed-in-place

_____ 7. Blanket and batt insulation is usually made from ___-resistant fiberglass or rock wool.
    A. fire
    B. moisture
    C. vermin
    D. all of the above

_____ **8.** Rigid foam insulation is available in ___ panels.
    A. faced
    B. unfaced
    C. various-sized
    D. all of the above

_____ **9.** Batt lengths of flexible insulation are normally ___″ in length.
    A. 16
    B. 24
    C. 48
    D. 72

_____ **10.** Frequently used loose fill insulation materials include rock wool, fiberglass, and ___.
    A. aluminum
    B. cellulose
    C. polystyrene
    D. polyurethane

Identification—Sealing Door Bottoms

_____ **1.** Felt or rubber sweep

_____ **2.** Interlocking threshold

_____ **3.** Automatic door

_____ **4.** Vinyl bulb threshold

Identification—Door and Window Weatherstripping

_____ **1.** Adhesive-backed foam rubber strip

_____ **2.** Rolled vinyl

_____ **3.** Spring metal V-strip

_____ **4.** Wood-backed foam rubber strip

True-False

T   F   **1.** Interior and exterior applications of rigid foam insulation to foundations are equally effective in reducing heat flow to the outside of a building.

T   F   **2.** Loose fill insulation may be blown into place.

T   F   **3.** Thermal insulation reduces sound transmission.

T   F   **4.** Thermal insulation has low fire-resistance ratings.

T   F   **5.** Exposure to sunlight can reduce the effectiveness of rigid foam insulation.

T   F   **6.** Blanket and batt insulation is commercially available in thicknesses ranging from 3″ to 7″ or more.

T   F   **7.** Blanket and batt insulation always has a vapor barrier that should face the heated side of the wall.

T   F   **8.** Rigid foam insulation panels are normally ¼″ thick.

T   F   **9.** Rigid foam insulation has a higher R value per inch of thickness than any other type of insulation.

T   F   **10.** Masonry veneer should not be used as an exterior wall over nonstructural rigid insulation panels.

T   F   **11.** Roofs may be insulated with rigid foam insulation panels.

T   F   **12.** Rigid foam insulation should not be attached to foundations using powder-actuated fasteners.

T   F   **13.** Chemical foam insulation has a high R value.

T   F   **14.** Foamed-in-place insulation expands to fill cavities.

T   F   **15.** Faced insulation batts have vapor barriers on both sides.

T   F   **16.** Foam insulation should not be blown into cavities of brick walls because it will cause the mortar to decay.

T   F   **17.** Heat may be transmitted through door and window materials.

T   F   **18.** Caulking is the best method of sealing small cracks to prevent heat loss.

T   F   **19.** Vapor barrier faced insulation is stapled through the flanges to joists or studs.

T   F   **20.** Loose fill insulation should not be used in wall cavities.

Name _____ Date _____

Completion

_____ **1.** Acceptable sound transmission coefficient (STC) ratings for sound insulation range from approximately ___ to ___.

_____ **2.** An impact insulation class number below ___ is unacceptable for sound control.

_____ **3.** The intensity of sound is expressed in ___.

_____ **4.** Staggered-stud or ___ walls have greater resistance to sound transmission than conventionally studded walls.

_____ **5.** To improve sound control, ___ on opposite sides of a hallway should be staggered.

_____ **6.** High levels of sound in a room are caused by sound waves being ___ from wall, ceiling, and floor surfaces.

_____ **7.** ___ glass in windows helps reduce sound transmission.

_____ **8.** Acoustical ___ is generally sprayed on ceilings.

_____ **9.** Smooth, hard building surfaces reflect up to ___% of the sound waves that strike them.

_____ **10.** ___ tile applied on a ceiling serves as an excellent sound absorber.

True-False

T   F   **1.** A low sound transmission class number on sound insulation indicates that little sound can pass through.

T   F   **2.** Airborne sound may be transmitted through a wall cavity and into an adjoining room by causing wall vibrations.

T   F   **3.** Many sound insulation materials are similar to thermal insulation materials.

T   F   **4.** Structure-borne sound transmission may be caused by heavy kitchen appliances.

T   F   **5.** Walking on a floor can cause impact sound transmission.

T   F   **6.** An impact insulation class number of 55 or more is required in a room for maximum privacy.

T   F   **7.** Light materials are more effective than heavy materials for blocking sound.

T   F   **8.** An interior wood stud wall with ½″ gypsum wallboard on each side is effective against loud speech.

T   F   **9.** Increased mass of an interior wall does not noticeably affect sound transmission through the wall.

T   F   **10.** Cavity absorption is a means of reducing sound transmission through a wall.

T   F   **11.** Breaking the sound vibration path of a wall aids in reducing sound transmission through the wall.

T   F   **12.** Carpets and carpet padding should be used in conjunction with approved floor-ceiling sound control methods to effectively reduce noise through overhead floors.

T   F   **13.** A double-stud wall has a higher STC rating than a staggered-stud wall.

T   F   **14.** Wood is more resilient than metal in the transmission of sound.

T   F   **15.** Noise from footsteps and vibrations from overhead floors are generally considered to be among the most disturbing household noises.

T   F   **16.** Walls constructed of metal studs and drywall transmit less sound vibration than walls constructed of wood studs and drywall.

T   F   **17.** Joint compound and tape applied to drywall has minimal effect on the reduction of sound transmission.

T   F   **18.** Solid wood-core doors are more effective in reducing sound transmission than hollow-core doors.

T   F   **19.** Sound striking an acoustical tile ceiling is quickly reflected, thereby reducing the sound level.

T   F   **20.** Acoustical tiles are commercially available in various sizes of square or rectangular shapes.

T   F   **21.** The printreading symbol for grade line is GL.

Printreading Symbols

This electrical symbol shows a(n) ___ outlet (light).

Solar Energy

Name _____ Date _____

Matching

_____ **1.** Passive solar heating

_____ **2.** Active solar heating

_____ **3.** Collector

_____ **4.** Absorber

_____ **5.** Conduction, convection, and radiation

_____ **6.** Direct gain

_____ **7.** Solarium

_____ **8.** Auxiliary heating system

_____ **9.** Storage

_____ **10.** Plenum

A. glass or plastic areas allowing sunlight into a building

B. simplest passive heating method

C. separate space to collect solar radiation

D. solar heating method with mechanical means

E. hard, darkened surface of storage material

F. needed for extended periods when sunlight is not available

G. open area at the bottom of a thermal storage unit

H. masonry products or water

I. solar heating method without mechanical means

J. methods of heat distribution

True-False

T   F   **1.** The front wall of a solar-heated house must face south.

T   F   **2.** Absorbers are hard, light-colored surfaces designed to attract heat.

T   F   **3.** Mechanical devices such as fans, ducts, and blowers are sometimes used to help circulate heat in a building with passive solar heating.

T   F   **4.** All solar heating methods require sunlight for efficient operation.

T   F   **5.** A window drape is an example of a control element used to prevent heat loss through collectors at night.

T   F   **6.** Masonry walls and floors of buildings utilizing a direct gain passive solar heating system should be at least 4″ thick.

T   F   **7.** The Thrombe wall system utilizes water-filled containers to store heat.

T   F   **8.** A time-lag heating process cannot take place in a solar-heated building that uses masonry walls for heat storage.

T   F   **9.** Water walls can absorb and store more heat than masonry walls of equal volume.

T   F   **10.** Active solar heating systems routinely provide 100% of the heating requirements of typical dwellings.

T   F   **11.** Active solar heating systems utilize mechanical means.

T   F   **12.** Active solar heating systems can be installed in new or existing buildings.

T   F   **13.** Collectors may be placed on the roof of a building or on a remote structure.

T   F   **14.** Auxiliary heating systems for buildings with solar heat should be natural gas, propane, or fuel oil, but not electricity.

T   F   **15.** The printreading symbol for ground is GND.

Short Answer

1. What is the primary difference between passive and active solar heating?

_____

_____

_____

2. Briefly describe the solar heating method.

_____

_____

_____

3. What five elements are required for a passive solar heating system?

_____

_____

_____

_____

4. How does an absorber differ from a collector in a passive solar heating system?

_____

_____

_____

5. What three methods of heat distribution are utilized in a passive solar heating system?

_____

_____

6. How does the Thrombe wall system differ from the water wall indirect gain method of passive solar heating?

_____

_____

_____

_____

7. What are the three main types of passive solar heating?

_____

_____

8. What five elements are required for an active solar heating system?

_____

_____

_____

9. Which passive solar heating system requires a separate space to capture solar heat and what type of separate space is used for this purpose?

_____

_____

_____

_____

10. What are the four operating modes of a fully automatic control system in a building heated by active solar heat with an auxiliary heating system?

_____

_____

_____

Identification—Solar Air Heating System

_____ **1.** High efficiency air cleaner

_____ **2.** Solar collector

_____ **3.** Storage unit

_____ **4.** Auxiliary heater

_____ **5.** Damper control

_____ **6.** Air handler

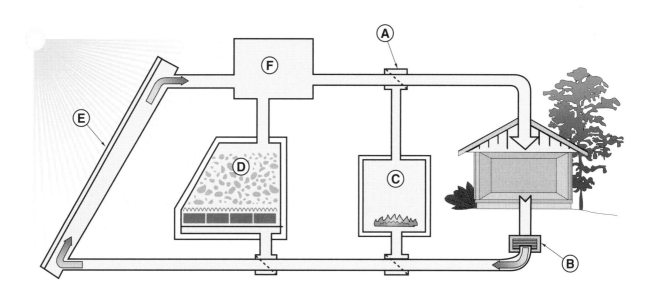

Name _____ Date _____

Matching

_____ 1. Headlap

_____ 2. Sidelap

_____ 3. Toplap

_____ 4. Shingle width

_____ 5. Exposure

_____ 6. Sheathing

_____ 7. Underlayment

_____ 8. Shingle

_____ 9. Shake

_____ 10. Flashing

A. distance that one shingle overlaps a shingle in the course below it

B. total measurement across the top of a shingle   .

C. roof covering split from a cedar log

D. plywood or nonveneered panel under roof shingle

E. asphalt-saturated felt placed over sheathing

F. distance between the exposed edges of overlapping shingles

G. distance that one shingle overlaps a shingle two courses below it

H. felt, metal, or plastic used to prevent water leakage where roofs intersect walls and around roof projections

I. roof covering made of asphalt or cedar

J. distance that one shingle overlaps the shingle next to it in the same course

Completion

_____ 1. The printreading abbreviation GYP BD stands for ___.

_____ 2. The cornice is located directly beneath the roof ___.

_____ 3. ___ are roof coverings that are sawed from a log.

_____ 4. The roof area must be ___ with plywood, OSB, or other nonveneered panels before roofing material is applied.

_____   **5.** Underlayment sidelap should not be less than ___″.

_____   **6.** ___ roof coverings are usually used on flat decks.

_____   **7.** ___ shingles are the most commonly used shingles on houses in the United States.

_____   **8.** Flashing may be used along the ___ of roofs in cold-weather areas to help protect the roof from damage due to ice buildup.

_____   **9.** Cedar is used for shingles and shakes because of its strength-to-weight ratio and because it does not ___ or ___ a large amount from moisture content.

_____   **10.** Metal flashing used with wood shingle or shake roofs should be ___ gauge or heavier.

_____   **11.** Standard lengths for wood shingles are ___″, ___″, and ___″.

_____   **12.** Cornices may be ___ or closed.

_____   **13.** While most spaces between rafters are blocked off, some may be left open and screened to provide ___ ventilation.

_____   **14.** ___ is one of the oldest types of finish covering used on pitched roofs.

_____   **15.** ___ must be toenailed to the wall and facenailed to rafter ends in a flat box cornice.

_____   **16.** ___ are roof coverings that are split from a log.

_____   **17.** Underlayment toplap should not be less than ___″.

_____   **18.** Clay tile is lighter than ___ tile.

_____   **19.** The ___ of a building are the portions of the roof that extend beyond the side walls.

_____   **20.** The trim piece nailed to the ends of rafters to finish off the roof edge is the ___ board.

_____   **21.** Three-way ___ tiles are used to cap the intersection of the end of the roof ridge and hip.

_____   **22.** Shakes are commercially available in standard lengths of ___″, ___″, and ___″.

_____   **23.** ___ on succeeding courses of asphalt shingles should be staggered.

_____   **24.** ___ is required where a roof meets a wall and around stacks projecting through the roof.

_____   **25.** Direct contact between shingles and sheathing is prevented by the ___.

_____   **26.** ___-colored shingles absorb heat more readily than ___-colored shingles.

_____ **27.** Hot-dipped, ___-coated nails are often used to nail wood shingles or shakes.

_____ **28.** Asphalt shingles have a life expectancy of ___ to ___ years.

_____ **29.** The ___ course of shakes on a roof should be doubled.

_____ **30.** Adjacent shakes should have a minimum spacing of ___ to allow for expansion.

True-False

T   F   **1.** Wide eaves of a dwelling help protect side walls from rain and snow.

T   F   **2.** Spaced sheathing should never be used as a base for nailing wood shingles or shakes.

T   F   **3.** The surface of shingles is smoother than the surface of shakes.

T   F   **4.** Underlayment should always be used on roofs that have asphalt shingles.

T   F   **5.** Roll tile is flat in cross section.

T   F   **6.** Roof overhangs are also known as eaves.

T   F   **7.** Sheathing is applied before underlayment is applied.

T   F   **8.** Field tiles are placed left to right.

T   F   **9.** Gutters may be made of wood, aluminum, plastic, or galvanized steel.

T   F   **10.** Built-up roof coverings are normally applied by carpenters.

T   F   **11.** Shingles give a more rustic appearance to the exterior of a building than shakes.

T   F   **12.** A minimum of three nails should be used to secure each wood shingle.

T   F   **13.** Nails for wood shingles or shakes should penetrate the sheathing at least ½″.

T   F   **14.** Shakes may be made of cedar, asphalt, mineral fiber, or fiberglass.

T   F   **15.** Lookouts provide a nailing base for the soffit.

T   F   **16.** The flat box and sloped box cornices are the most common types of closed cornices.

T   F   **17.** A frieze board should never be notched between rafters.

T   F   **18.** The open valley method is considered the most practical method of finishing off a valley.

T   F   **19.** Counterflashing around a chimney should extend ¾″ into the mortar joints of the chimney.

T   F   **20.** Soffits are nailed to the ends of rafters.

T	F	**21.** Gutters and downspouts are attached to the frieze board to provide water runoff from the roof.
T	F	**22.** Wood shingles and shakes are produced from cedar logs.
T	F	**23.** The soffit is nailed directly to the lookouts in a sloped box cornice.
T	F	**24.** Underlayment is not required for roofs covered with wood shingles.
T	F	**25.** Asphalt strip shingles are preferred for new construction over individual shingles.
T	F	**26.** Felt underlayment is not required for roofs with asphalt shingles.
T	F	**27.** Mineral-surfaced roofing material is placed face down in the valley in closed valley flashing.
T	F	**28.** Staples are commonly used to fasten asphalt shingles.
T	F	**29.** Local fire codes may prohibit the use of wood shingles or shakes.
T	F	**30.** Straightsplit shakes have no taper.
T	F	**31.** Wood shingles are commercially available in random widths and standard lengths.
T	F	**32.** Joints between shakes are offset from adjacent courses.
T	F	**33.** A strip of underlayment should be placed between each course of shakes as each course is nailed.
T	F	**34.** Shakes are longer than wood shingles.
T	F	**35.** Shingles and shakes should not be applied to roofs having a unit rise of less than ___.

Identification—Shingles

_____ **1.** American

_____ **2.** Hip or ridge

_____ **3.** Three-tab hexagonal

_____ **4.** Interlocking

_____ **5.** Dutch

_____ **6.** Two-tab hexagonal

_____ **7.** Two-tab shingle

_____ **8.** Three-tab shingle

Name _____ Date _____

True-False

T   F   **1.** Openings for natural ventilation are not required if a complete mechanical ventilation system exists in the structure.

T   F   **2.** Fixed-sash windows allow more ventilation than double-hung windows.

T   F   **3.** The minimum glass and venting area for inhabited rooms may be specified by local building codes.

T   F   **4.** Double-hung windows have two bottom sashes.

T   F   **5.** A window frame should be equal to the thickness of the rough wall in which it is installed.

T   F   **6.** Window and door units are set in place by carpenters.

T   F   **7.** The tops of windows normally line up with the tops of doors in a building.

T   F   **8.** Horizontal sliding windows operate on tracks located on either side of the window.

T   F   **9.** Wood is the oldest type of material used for window units.

T   F   **10.** Casement windows may be installed to swing out of or into a building.

T   F   **11.** Awning windows may be combined with fixed-sash windows in the same opening.

T   F   **12.** Hopper windows have a series of small glass lights set into metal clips on each side of the frame.

T   F   **13.** Horizontal sliding window units may have fixed and movable windows.

T   F   **14.** All four sides of a rough window opening should be flashed with building paper to provide an additional weatherseal to the window unit.

T   F   **15.** Rafters should never be cut when installing skylights.

T   F   **16.** Hopper windows may swing in or out.

T   F   **17.** Awning windows may swing in or out.

T   F   **18.** Overhead garage doors are the most common type of garage door.

T   F   **19.** A sectional roll-up garage door is less expensive than a one-piece swing-up garage door.

T   F   **20.** The printreading abbreviation for hardwood is HDW.

Identification—Typical Overhead Garage Door Hardware

_____   **1.** Torsion spring

_____   **2.** Weatherstripping

_____   **3.** Rollers

_____   **4.** Angled closing

_____   **5.** Hinges

_____   **6.** Counterbalance

_____   **7.** Door latch

_____   **8.** Transmitter

_____   **9.** Electric motor unit

_____   **10.** Reinforced strut

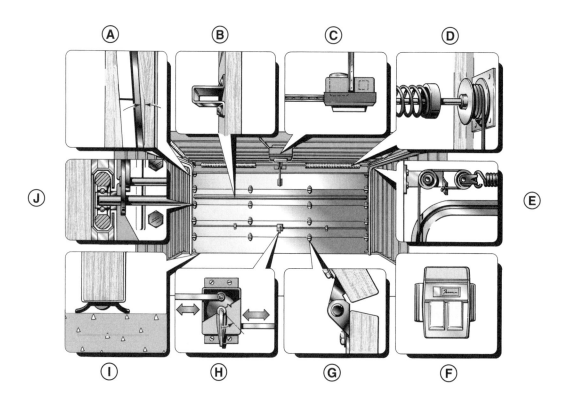

Multiple Choice

_____ **1.** Information concerning door and window units of a building is found on the ___ of the prints.
   A. floor plans
   B. elevation plans
   C. details
   D. all of the above

_____ **2.** Ventilation through a double-hung window cannot exceed ___% of the window opening.
   A. 20
   B. 25
   C. 50
   D. 75

_____ **3.** Each pane of glass in a window with muntins is known as a ___.
   A. unipane
   B. sheet
   C. light
   D. none of the above

_____ **4.** Which of the following is not one of the three basic types of overhead garage doors?
   A. sectional roll-up door
   B. one-piece swing-up door
   C. sliding pocket door
   D. rolling steel door

_____ **5.** Sliding glass doors are often referred to as ___ doors.
   A. porch
   B. patio
   C. interior
   D. none of the above

_____ **6.** ___ may be used to hold the sashes of a double-hung window in an open vertical position?
   A. Springs
   B. Compressible weatherstripping
   C. Sash weight and rope
   D. all of the above

_____ **7.** A window frame is composed of ___.
   A. one top piece and two side pieces
   B. two top pieces and one side piece
   C. two top pieces and two side pieces
   D. none of the above

_____ **8.** Casement windows are hinged on ___.
    A. one side
    B. two sides
    C. either side and the top
    D. two sides and the bottom

_____ **9.** Hopper windows are hinged at the ___ and swing out at the ___.
    A. top; bottom
    B. top; side
    C. bottom; top
    D. side; top

_____ **10.** Space between the frame of a window and the wall is covered by the ___.
    A. apron
    B. sill
    C. casing
    D. sash

_____ **11.** Casement windows allow ___% ventilation through the window opening.
    A. 25
    B. 50
    C. 75
    D. 100

_____ **12.** The framework, either wood or metal, that holds the glass is the window ___.
    A. light
    B. muntin
    C. casing
    D. none of the above

_____ **13.** Awning windows are hinged at the ___ and swing out at the ___.
    A. top; bottom
    B. top; side
    C. bottom; top
    D. side; top

_____ **14.** The lights of a jalousie window may be opened to ___°.
    A. 45
    B. 90
    C. 100
    D. 180

_____ **15.** The slanted piece at the bottom of a window frame is the ___.
    A. muntin
    B. sash
    C. jamb
    D. sill

Identification—Double-Hung Window Components

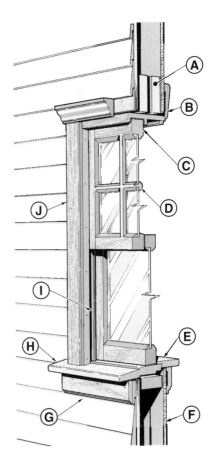

_____ **1.** Muntin

_____ **2.** Rough sill

_____ **3.** Double header

_____ **4.** Side jamb

_____ **5.** Exterior apron

_____ **6.** Upper sash top rail

_____ **7.** Stool

_____ **8.** Exterior side casing

_____ **9.** Exterior sill

_____ **10.** Interior top casing

Completion

_____ **1.** Some manufacturers recommend that the window sash be removed before setting the ___ of a window unit to avoid possible damage to the sash.

_____ **2.** Skylights may be installed over a light ___ in an attic roof.

_____ **3.** ___ doors provide passage between the inside and outside of a building.

_____ **4.** ___ and ___ window units are easier to install than wood window units.

_____ **5.** ___ are placed in the roof of a building to allow light to enter living space in the building.

_____ **6.** ___ garage doors are available for single-car or double-car garages.

_____ **7.** Rolling steel doors are normally used in industrial or ___ buildings.

_____ **8.** ___ units and exterior ___ units are usually assembled at a factory or mill-cabinet shop.

Printreading Symbols

This electrical symbol shows a(n) ___.

M

_____

Name _____   Date _____

Identification—Board Siding Patterns

_____ **1.** Bevel

_____ **2.** Log cabin

_____ **3.** Drop

_____ **4.** Dolly Varden

_____ **5.** Board

_____ **6.** Tongue-and-groove

_____ **7.** Channel rustic

_____ **8.** Bungalow

Ⓐ   Ⓑ   Ⓒ

Ⓓ   Ⓔ   Ⓕ

Ⓖ   Ⓗ

True-False

T   F   **1.** Wood and masonry materials should not be combined for the exterior wall finish of a building.

T   F   **2.** Hardboard may be painted for appearance, but it should not be stained.

T   F   **3.** In the southwestern states of the U.S., moisture content of wood board siding between 8% and 9% is acceptable.

T   F   **4.** Triple-lap board siding requires longer installation time than common board siding.

T   F   **5.** Hardboard can be manufactured to appear like many different lumber species.

T   F   **6.** Board siding is usually applied in a vertical position.

T   F   **7.** Building paper is water resistant.

T   F   **8.** Siding should always be applied over sheathing nailed to the wall studs.

T   F   **9.** Siding must be nailed through the sheathing and into the wall studs when nonstructural sheathing is used.

T   F   **10.** The ends of siding should be mitered for a tight fit.

T   F   **11.** A drip cap aids in preventing water infiltration where vertical and horizontal siding meet.

T   F   **12.** Siding placed vertically on walls not covered with structural sheathing requires nailing blocks.

T   F   **13.** Casing nails do not provide proper holding power for board siding.

T   F   **14.** Panel siding can be installed more quickly than board siding.

T   F   **15.** Plywood panels with structural ratings do not require diagonal bracing inside the framed wall.

T   F   **16.** Pneumatic nailers should not be used to fasten ½″ siding.

T   F   **17.** Hardboard panels have greater shear rack resistance than plywood panels.

T   F   **18.** Nails should be countersunk for siding that will be painted.

T   F   **19.** Wood shingles and shakes used for roof covering may also be applied to walls.

T   F   **20.** Pneumatically driven staples are not recommended when installing wood shingles on exterior walls.

T   F   **21.** Panel systems for shingles provide panels with a number of shingles bonded to the panel.

T   F   **22.** Shingles should not be nailed directly to nonstructural sheathing.

T   F   **23.** Aluminum siding may be installed by carpenters.

T   F   **24.** Some building codes may require that aluminum siding be grounded.

T   F   **25.** Joints between lengths of aluminum siding should be vertically aligned.

T   F   **26.** The vertical placement of the header determines the height of the deck floor.

T   F   **27.** A water table may be placed below the bottom course of siding to divert water drainage from the foundation wall.

T   F   **28.** Aluminum siding should never be applied directly to sheathing.

T   F   **29.** Corners of horizontal board siding may be mitered.

T   F   **30.** Aluminum siding is produced with or without insulating backing board.

T   F   **31.** Butt joints of board siding should be staggered.

T   F   **32.** Edges of plywood panels should be sealed at the time of installation.

T   F   **33.** When double coursing shingles, the under course should be applied using higher grade shingles.

T   F   **34.** Aluminum siding is resistant to corrosion.

T   F   **35.** Joists for decks may be placed 16″ or 24″ OC.

Identification—Corner Finishes

_____   **1.** Mitered corners

_____   **2.** Butt joint corners

_____   **3.** Corner pieces over siding

_____   **4.** Inside corner strips

Ⓐ   Ⓑ   Ⓒ   Ⓓ

Multiple Choice

_____   **1.** ___ is commonly used for siding.
  A. Wood
  B. Vinyl
  C. Aluminum
  D. all of the above

_____   **2.** Flashing material used for exterior walls is usually a(n) ___ material.
  A. galvanized or vinyl
  B. asphalt or cementitious
  C. bituminous or cementitious
  D. none of the above

_____   **3.** Wood and wood-fabricated siding is available in ___.
  A. panels
  B. boards
  C. shingles or shakes
  D. all of the above

_____ **4.** Allowable moisture content for wood board siding in most areas of the U.S. is ___% to ___%.

    A. 8; 12

    B. 10; 12

    C. 12; 14

    D. 14; 16

_____ **5.** Board siding may be applied in a(n) ___ position.

    A. vertical

    B. horizontal

    C. diagonal

    D. all of the above

_____ **6.** Redwood and cedar are highly recommended for wood board siding because they are ___.

    A. readily available in all parts of the U.S.

    B. very inexpensive

    C. highly resistant to decay

    D. none of the above

_____ **7.** When applying board siding in a horizontal position, the ___ row of boards should be applied first.

    A. top

    B. bottom

    C. third

    D. none of the above

_____ **8.** ___ nails are recommended for use with horizontal and vertical wood siding.

    A. Aluminum alloy

    B. Stainless steel

    C. Hot-dipped galvanized

    D. all of the above

_____ **9.** The bottom row of boards, when applying board siding horizontally, should extend at least ___″ below the top of the foundation wall.

    A. 1

    B. 3

    C. 6

    D. 12

_____ **10.** Plywood panels may be applied ___.

    A. to studs placed 16″ or 24″ OC

    B. over nonstructural sheathing

    C. either vertically or horizontally

    D. all of the above

Completion

_____   1. Wood board siding should be ___ with a sealer before nailing.

_____   2. ___ should be applied over drip caps of doors and windows to prevent rain leakage.

_____   3. Siding may be nailed directly into ___ sheathing without penetrating wall studs.

_____   4. Standard lengths of plywood panels include 7′, 8′, 9′, 10′, and ___′ lengths.

_____   5. Hardboard panels are usually ___″ thick.

_____   6. A wood rod, known as a(n) ___, may be used when laying out board siding.

_____   7. Hardboard panels should be stored on the job site at least ___ days before installation.

_____   8. When nailing ½″ board siding, ___d nails are commonly used.

_____   9. Wood shingles applied to walls should have a(n) ___″ to ___″ space between vertical edges.

_____   10. ___ nails should be used when building wood porches and decks to prevent streaking.

_____   11. Panels are normally placed with their long sides in a(n) ___ position.

_____   12. Wood shingle panel systems are commercially available in ___′ and ___′ widths.

_____   13. For panels over ½″ thick, ___d nails should be used.

_____   14. The printreading abbreviation HWH stands for ___.

Name _____ Date _____

True-False

T **F**   **1.** Predecorated gypsum board panels are commercially available for installation in new residential construction.

T **F**   **2.** Gypsum board should be cut with a coarse-tooth handsaw or an electric handsaw.

T **F**   **3.** The double-ply system of installing wallboard provides greater sound transmission than the single-ply system.

T F   **4.** Joints in drywall should be staggered when possible.

T F   **5.** Most new homes are constructed with gypsum board wall finish.

T F   **6.** Drywall may be applied with screws.

T F   **7.** Laminating and stud adhesives may be used to fasten gypsum board directly to concrete or masonry.

T F   **8.** Standard sizes of readily available plywood paneling are 4′ × 8′ and 4′ × 12′.

T F   **9.** Predecorated gypsum board requires no additional finish during installation.

T F   **10.** Furring strips, to which paneling may be nailed, are usually 1″ × 2″ in size.

T **F**   **11.** Hardboard paneling does not require preconditioning before application.

T **F**   **12.** Nails, adhesives, or screws may be used to fasten wall paneling.

T **F**   **13.** Gypsum board has no sound-insulating qualities.

T F   **14.** Plywood paneling is commercially available with plain or textured surfaces.

T **F**   **15.** A resawn finish on a solid board panel is smooth.

T F   **16.** Tongue-and-groove boards are blind-nailed at a 45° angle.

T F   **17.** Carpenters may hang gypsum board.

T F   **18.** Adhesives are not suitable for installing gypsum board on interior walls.

T F   **19.** The majority of plywood paneling sold today is prefinished.

T F   **20.** Plywood paneling should be stored on the job site for several days prior to application.

T F **21.** Gypsum board has good fire-resistant qualities.

T F **22.** Gypsum board cannot be installed over metal stud walls.

T F **23.** Predecorated gypsum board panels are usually fastened with drywall screws.

T F **24.** Preconditioning of plywood paneling eliminates significant shrinkage after the panels are applied.

T F **25.** Board-on-board and board-and-batten panels should be applied horizontally.

Printreading Symbols

Identify the type of opening shown in this plan view.

window

Completion

vertical   **1.** The ___ system of installing wallboard is most often used in residential construction.

_1/16"_   **2.** Nails used to install drywall should penetrate at least ___″ into the wood.

tapers   **3.** Tradesworkers called ___ specialize in finishing gypsum board panels.

ceilings   **4.** When applying drywall to walls and ceilings, the panels should be applied to the ___ first.

metal   **5.** ___ corner beads are used to reinforce outside corners of gypsum board walls.

single double   **6.** Gypsum board may be nailed using either the ___ or ___ method.

compound   **7.** Joint ___ is used to cover taped joints and nail dimples.

face   **8.** Most plywood wall panels have a hardwood ___, such as oak, birch, or mahogany.

_8 , 12_   **9.** For plywood paneling, nails should be placed ___″ OC along the edges and ___″ OC at intermediate studs or furring strips.

base   **10.** The bottom of paneling is covered with ___ molding.

_1/4"_   **11.** Hardboard paneling is commercially available in thicknesses from ⅛″ to ___″.

scribing   **12.** Paneling that meets in a corner may require ___ to obtain a close fit when molding is not applied.

_____ plumb **13.** When applying paneling to a wall the first piece of paneling must be ___.

_____ 4' 12 **14.** Boards for solid board paneling are usually ___″ to ___″ wide.

_____ **15.** For hardboard paneling, nails should be placed ___″ OC along the edges and ___″ OC at intermediate studs or furring strips.

_____ **16.** Standard gypsum board is commercially available in ___′ widths.

Multiple Choice

_____ **1.** Gypsum board can be fastened to ___.
A. wood
B. metal
C. concrete
D. all of the above

_____ **2.** When a single layer of drywall is installed, the long edge can be applied ___.
A. horizontally
B. vertically
C. with the studs
D. all of the above

_____ **3.** The rough ceiling height in residential construction is usually ___′-___″.
A. 8; 0
B. 8; 1
C. 9; 0
D. none of the above

_____ **4.** Drywall nails should be driven ___.
A. flush with the surface
B. slightly above the surface
C. slightly below the surface to cause a dimple
D. none of the above

_____ **5.** When installing tongue-and-groove paneling, ___ eliminates the need to countersink and putty the face nails.
A. moisture control
B. a furring strip
C. blind nailing
D. none of the above

_____ **6.** When solid boards are fastened ___ to a stud wall, blocking must be placed between the studs.
A. vertically
B. horizontally
C. diagonally
D. all of the above

_____ **7.** Ceiling tiles used in residential and commercial construction are ___.
      A. fabricated from fiberboard, mineral glass, etc.
      B. available in various colors and designs
      C. available with acoustical qualities
      D. all of the above

_____ **8.** Plastic laminate premounted on plywood or particleboard panels is commercially available in widths of ___″ and lengths of ___′.
      A. 6 to 12; 2 to 4
      B. 12 to 24; 4 to 6
      C. 16 to 48; 8 to 10
      D. all of the above

_____ **9.** ___, at right angles to each other, must be located on the ceiling before the first row of tiles is installed directly on a ceiling.
      A. Furring strips
      B. Reference points
      C. Centerlines
      D. none of the above

_____ **10.** The printreading abbreviation for insulation is ___.
      A. INSL
      B. ISN
      C. INS
      D. none of the above

Identification—Suspended Ceiling Components

_____ **1.** Main runner

_____ **2.** Hanger wire

_____ **3.** Cross tee

_____ **4.** Outer trim

Printreading Symbols

Identify the type of opening shown in this wall.

_____

Name _____ Date _____

True-False

T   F   **1.** Solid-core doors are generally less expensive than hollow-core doors.

T   F   **2.** Waterproof adhesive may be used for interior doors.

T   F   **3.** Solid-core doors provide better sound insulation than hollow-core doors.

T   F   **4.** Jamb length of door assemblies is precut to the exact height during manufacture and assembly of the unit.

T   F   **5.** A properly fitted door has equal clearance on the top and the two sides.

T   F   **6.** Solid-core doors are lighter than hollow-core doors.

T   F   **7.** Hollow-core doors are not recommended for exterior door use.

T   F   **8.** A rabbeted jamb does not require a door stop.

T   F   **9.** Split-jamb door units are not adjustable for different wall widths.

T   F   **10.** Bi-fold doors are usually 1⅛″ or 1⅜″ thick.

T   F   **11.** Solid-core doors are not available with fire ratings.

T   F   **12.** Pocket sliding doors are also known as recessed doors.

T   F   **13.** Multifolding doors are supported by roller hangers that run along an overhead track.

T   F   **14.** A right-hand door hinges on the right and opens outward.

T   F   **15.** An escutcheon is a trim piece of a lock set.

T   F   **16.** The second door of a bi-fold door is hinged to the first door.

T   F   **17.** Metal door frames do not usually require trimming.

T   F   **18.** Many locksets are reversible so that they may be adjusted to operate in doors that swing in either direction.

T   F   **19.** Steel doors are widely used in commercial construction.

T   F   **20.** Holes must be drilled in the door to install cylindrical locks but not tubular locks.

T F **21.** Dead bolts are usually keyed on the outside of the door and have a knob or handle on the inside.

T F **22.** Miter joints or butt joints may be used with door stops.

T F **23.** For doors that open between rooms, the keyed side of the lock is considered the inside of the door.

T F **24.** A double-acting floor hinge allows a door to swing in two directions.

T F **25.** The printreading symbol for interior is INT.

Multiple Choice

_____ **1.** The hinge size required to hang a door depends on the ___ of the door.
  A. weight
  B. thickness
  C. width
  D. all of the above

_____ **2.** The vertical members of a panel door are known as ___.
  A. rails
  B. stiles
  C. faceplates
  D. none of the above

_____ **3.** Information concerning doors that can be determined from the prints includes ___.
  A. width of doors
  B. the direction in which doors swing
  C. height of doors
  D. all of the above

_____ **4.** Door jambs should be set ___″ wider than the door width.
  A. $\frac{1}{16}$
  B. $\frac{1}{8}$
  C. $\frac{3}{16}$
  D. $\frac{1}{4}$

_____ **5.** Normally, the upper hinge of a door is placed ___″ from the top of the door.
  A. 5
  B. 10
  C. 12
  D. none of the above

_____ **6.** In a prehung door unit, the ___.
    A. door is hinged in the frame
    B. hole is predrilled for the lock
    C. door stops are tacked to the jamb
    D. all of the above

_____ **7.** Steel doors are commercially available in thicknesses of ___″ and ___″.
    A. 1⅜; 1½
    B. 1⅜; 1⅝
    C. 1⅜; 1¾
    D. 1⅜; 1⅞

_____ **8.** A left-hand reverse door ___.
    A. hinges on the left and opens inward
    B. hinges on the left and opens outward
    C. hinges on the right and opens inward
    D. hinges on the right and opens outward

_____ **9.** Door stops are designed primarily to ___.
    A. keep the door in an open position
    B. keep the door in a closed position
    C. protect the wall
    D. prevent the lock from being set accidentally

_____ **10.** A right-hand reverse door hinges on the ___.
    A. right and opens inward
    B. right and opens outward
    C. left and opens inward
    D. left and opens outward

Identification—Door Hand

_____ **1.** Left-hand
_____ **2.** Left-hand reverse
_____ **3.** Right-hand
_____ **4.** Right-hand reverse

Completion

_____ **1.** Interior doors are usually ___″ thick.

_____ **2.** ___ is the trim placed around a doorjamb.

_____ **3.** Two main types of flush doors are ___ and ___.

_____ **4.** The ___ is the finish frame of a door unit in which the door hangs.

_____ **5.** Panel doors may have ___ or raised panels.

_____ **6.** A jamb assembly includes one head jamb and ___ side jambs.

_____ **7.** ___ should be used to plumb and align the side pieces of a door jamb during installation.

_____ **8.** Exterior doors are usually ___″ thick.

_____ **9.** A door should be ___ on the lock side to prevent the inside edge from scraping against the jamb as the door is closed.

_____ **10.** The ___ is the part of the lock that contains the keyhole and tumbler mechanism.

_____ **11.** Heavy doors and doors over ___ in height require a third hinge centered between the top and bottom hinges.

_____ **12.** Double doors often have a(n) ___ door that is held in place by flush bolts.

_____ **13.** ___ doors are often used when swinging doors are impractical.

_____ **14.** The door ___ is the direction in which a door swings.

_____ **15.** Panel doors are also known as ___ doors.

_____ **16.** ___ drawings of a print usually provide information about the jamb, casing, and stop material to be used with doors.

_____ **17.** In residential construction, the height of the lock from the floor to its center is usually ___″.

_____ **18.** ___ spring hinges are often used on doors in public buildings to automatically close a door after it has been opened.

Printreading Symbols

What type of outlet is shown? _____

Name _____   Date _____

Identification—Kitchen Base Cabinets

_____   **1.** Toeboard

_____   **2.** Cleat

_____   **3.** Mounting rail

_____   **4.** Face frame

_____   **5.** Stile

_____   **6.** Mullion

_____   **7.** Web frame

_____   **8.** Center drawer slide

Completion

_____   **1.** The ___ frame of a cabinet fits on the front of the cabinet.

_____   **2.** In residential construction, most cabinets and countertops are located in the ___ and ___.

_____   **3.** Rails and stiles of a cabinet face frame may be joined with ___, ___, or ___ joints.

_____   **4.** A(n) ___ may be utilized to prevent wall cabinets from tipping forward or falling during installation.

_____   **5.** When hanging wall cabinets, wood screws should penetrate through the back of the cabinet and into the wall ___.

_____   **6.** Wall cabinets should be installed so they are ___ and ___ on the wall.

_____   **7.** ___ screws are generally used to attach knobs and pulls on cabinet doors and drawers.

_____ **8.** The rear vertical portion of a countertop is known as a ___.

_____ **9.** The two major types of cabinets used for kitchens are ___ and ___ cabinets.

_____ **10.** Lipped doors on cabinets are simpler to install than ___ doors.

True-False

T   F   **1.** Side-mounted drawer slides can usually support more weight than bottom-mounted drawer slides.

T   F   **2.** Wall cabinets may, in some instances, be fastened together prior to being hung on the wall.

T   F   **3.** Plastic laminate should be bonded to its base with white or yellow glue.

T   F   **4.** Sink openings in countertops may be made on the job.

T   F   **5.** Wood drawer guides may stick due to low humidity.

T   F   **6.** Cabinet backs are best attached by rabbeting the back edge of the cabinet sides.

T   F   **7.** Solid-surface countertops should be supported every 12″ for ½″ thicknesses.

T   F   **8.** Plastic laminate should not be cut with power tools.

T   F   **9.** Generally, more cabinets are required for residential construction than for commercial construction.

T   F   **10.** Plastic laminate should be cut larger than the base to which it will be applied and trimmed to size after application.

T   F   **11.** Wood drawer guides are seldom used in kitchen cabinets.

T   F   **12.** When applying plastic laminate to a base piece, apply edges before the top is applied.

T   F   **13.** Semi-concealed pivot hinges are often used with overlay doors.

T   F   **14.** Flush doors of a kitchen wall cabinet fit inside the opening of the cabinet.

T   F   **15.** An effective means of securing an island base cabinet to the floor is to nail a 2 × 4 to the floor and secure the cabinet toekick to the 2 × 4.

Name _____ Date _____

True-False

T   F   **1.** Plastic or metal trim is often used in commercial construction.

T   F   **2.** Wood molding is manufactured from hardwood lumber only.

T   F   **3.** Chair rails are primarily decorative.

T   F   **4.** Door casing should be backed out to provide a tight fit between the jamb and the wall.

T   F   **5.** A window stool has a miter joint at each end to make a corner return.

T   F   **6.** Base caps are usually used with modern-design molding.

T   F   **7.** Miter joints are not desirable for inside corners of base molding.

T   F   **8.** A scarf joint for base molding is made by overlapping two 45° angles.

T   F   **9.** Wood is the major material used for interior trim in residential construction.

T   F   **10.** Base molding is applied at the bottom of walls.

T   F   **11.** Ceiling molding is usually attached to the wall with 4d nails driven into the studs.

T   F   **12.** The window casing in a house should be the same design as the door casing.

T   F   **13.** Base shoe molding should be installed before applying base molding to a wall.

T   F   **14.** A contemporary window frame requires four pieces of casing.

T   F   **15.** Traditional windows require a rabbeted stool before the casing can be nailed.

Printreading Symbols

Does this air duct supply or return air?

_____

Completion

_____ 1. ___ is applied around doors and windows.

_____ 2. When nailing the narrow edge of tapered casing to a doorjamb, ___d or ___d nails should be used.

_____ 3. A(n) ___° miter is usually cut in door casing at the joint between top and side pieces.

_____ 4. A(n) ___ is a piece of resawn lumber that will produce a molding pattern with the least amount of waste.

_____ 5. Molding is usually fastened with ___ nails.

_____ 6. A(n) ___ is the space where the casing is held short on the jamb.

_____ 7. Joints in long runs of base molding should fall over a ___.

_____ 8. Base molding is installed after the ___ has been nailed in place.

_____ 9. Base molding is also known as ___.

_____ 10. ___ joints are recommended for inside corners of base molding.

Identification—Moldings

_____ 1. Base

_____ 2. Base shoe

_____ 3. Casing

_____ 4. Base cap

_____ 5. Handrail

_____ 6. Astragal

_____ 7. Chair rail

_____ 8. Cove

Name _____ Date _____

True-False

T	F	**1.** Vinyl and slate are examples of resilient tiles.
T	F	**2.** Wood flooring is commercially available in strips, planks, and blocks.
T	F	**3.** Resilient flooring may be in sheet form.
T	F	**4.** Hardwood flooring is more durable than softwood flooring.
T	F	**5.** Strip flooring should never be top nailed because of its tendency to split.
T	F	**6.** Subfloors are covered with asphalt-saturated felt or building paper before strip flooring is applied.
T	F	**7.** Strip flooring needs to adapt to local humidity before installation.
T	F	**8.** Screeds should never be nailed to a concrete floor because of the possibility of water seepage.
T	F	**9.** When nailing strip flooring, nails should be driven at a 45° to 50° angle.
T	F	**10.** Tongue-and-groove edges and ends are found on $^{25}/_{32}''$ thick plank flooring.
T	F	**11.** Parquet flooring is made of wood strips arranged in various patterns.
T	F	**12.** Laminated block flooring is an excellent choice for use in damp locations.
T	F	**13.** Resilient flooring is usually laid by carpenters.
T	F	**14.** When laying floor tiles, the first row should be laid so that it butts against the longest wall of the room.
T	F	**15.** Strip flooring with tongue-and-groove joints should be blind-nailed when possible.
T	F	**16.** Nails through strip flooring should penetrate into the floor joists if possible.
T	F	**17.** Plank flooring is commercially available in widths ranging from 3″ to 8″.
T	F	**18.** Underlayment panels for resilient floors may be glued, but they should not be nailed.
T	F	**19.** Resilient floor materials can be cut with a knife or a pair of scissors.

T   F   **20.** Hardwood flooring should not be placed by carpenters.

T   F   **21.** Strip flooring is available in lengths exceeding 16′.

T   F   **22.** Joists should run parallel to strip flooring whenever possible.

T   F   **23.** Screeds should be applied to a concrete subfloor before strip flooring is applied.

T   F   **24.** Most block flooring has tongue-and-groove edges to assure alignment between the squares.

T   F   **25.** Resilient floor materials are produced in tile and sheet form.

T   F   **26.** Wall-to-wall carpeting is considered a finish floor material.

T   F   **27.** Strip flooring is from 1½″ to 3½″ wide.

T   F   **28.** Plank flooring is usually used in buildings of traditional design.

T   F   **29.** Screws attached through the face of plank flooring should be counterbored and covered with wood plugs.

T   F   **30.** The printreading symbol for kitchen is KIT.

Matching

_____   **1.** Finish floor materials

_____   **2.** Softwood finish flooring

_____   **3.** End-matched

_____   **4.** Hollow-backed

_____   **5.** Screed

_____   **6.** Solid unit, laminated, and slat

_____   **7.** Resilient

_____   **8.** Hardwood finish flooring

_____   **9.** Subfloor

_____   **10.** Slat blocks

A. manufactured mostly of oak, beech, birch, or maple

B. recess in flooring, which allows it to lie flat

C. types of block flooring

D. nailing strip

E. mosaic or parquet flooring

F. able to yield under pressure

G. wood, tile, and carpeting

H. tongue-and-groove fit where pieces butt together

I. plywood, OSB, or performance-rated panels

J. manufactured mostly of southern pine, Douglas fir, and western hemlock

Name _____ Date _____

Completion

_____ **1.** ___ determines the stairway arrangement.

_____ **2.** ___ stairways are located inside a building.

_____ **3.** Wedge-shaped steps in an L-shaped stairway are known as ___.

_____ **4.** ___ stairways serve the uninhabited areas of a building.

_____ **5.** Balusters are also known as ___.

_____ **6.** The portion of a stairway that supports the treads is the ___.

_____ **7.** The horizontal distance measured from the foot of a stairway to the point where the stairway ends above is the ___.

_____ **8.** The exact tread and riser sizes for a stairway are based on the ___ and ___.

_____ **9.** ___ stairways are located outside a building.

_____ **10.** A(n) ___ pole may be used to determine riser heights for stairs.

_____ **11.** A stairway with 7½″ risers should have treads not exceeding ___″.

_____ **12.** ___ are positioned on a framing square to ensure consistent measurements during layout.

_____ **13.** When using a framing square to lay out treads and risers on a stringer, the tread width is read from the ___ of the framing square.

_____ **14.** ___ stairways, either circular or elliptical, are usually prefabricated in a shop.

_____ **15.** Finish treads and risers are nailed to the ___.

Printreading Symbols

Identify this wall symbol. _____

Matching

_____ **1.** Landing

_____ **2.** Closed stringer

_____ **3.** Riser

_____ **4.** Nosing

_____ **5.** Tread

_____ **6.** Open stringer

_____ **7.** Baluster

_____ **8.** Handrail

_____ **9.** Starting newel post

_____ **10.** Landing newel post

A. finish board nailed against wall of the stairway

B. tread projection beyond the face of the riser

C. cut-out piece supporting the open side of a stairway

D. main post supporting the handrail at the bottom of stairs

E. upright pieces running between the handrail and the treads

F. platform that breaks the stair flight between floors

G. the step on which the foot is placed

H. provides support when using stairway

I. main post supporting the handrail at landing

J. forms the vertical face of the step

True-False

T   F   **1.** A straight-flight stairway with no landing is the simplest stairway to construct.

T   F   **2.** Cove molding may be used on stairways only for covering the joint between the tread and the stringer.

T   F   **3.** An L-shaped stairway runs along two walls.

T   F   **4.** Prefabricated stairways usually have housed stringers.

T   F   **5.** No more than two stringers should be used for a straight-flight stairway.

T   F   **6.** The nosing return is the projection over the face of the stringer at the end of the tread.

T   F   **7.** A straight-flight stairway may turn halfway between floors.

T   F   **8.** Cut-out stringers are used only for interior stairways.

T   F   **9.** The preferred angle for a stairway is 45°.

T   F   **10.** Dividing the total rise by the number of risers gives the height of the riser.

T   F   **11.** U-shaped stairways have two sets of stairways that run parallel to one another.

T   F   **12.** The thickness of the tread material is a factor to consider when determining distance between stringers.

T   F   **13.** Critical angles are the most comfortable walking angles for a stairway.

T   F   **14.** Stringers provide the main support for a stairway.

T   F   **15.** The balustrade of a stairway is formed by the balusters and handrail.

T   F   **16.** Riser height is also known as total rise.

T   F   **17.** A riser height of 7⅜″ with a tread width of 10½″ falls within acceptable tread and riser combinations.

T   F   **18.** Width of the headers is not considered when determining the total rise.

T   F   **19.** There is always one riser more than the total number of treads in a stairway.

T   F   **20.** A shorter riser may be combined with a wider tread to form a preferred-angle stairway.

Multiple Choice

1. The minimum components required to form stairways are ___.
A. treads and stringers
B. landings and risers
C. handrails and balusters
D. all of the above

2. The landing of a stairway is the ___.
A. curved or bent section on the handrail
B. piece forming the vertical face of the step
C. platform that breaks the flight between floors
D. finish board nailed against the wall

3. Generally, stringers should be installed ___ OC when 1½″ treads are used.
A. 24
B. 30
C. 36
D. 42

4. The vertical distance from one floor to the floor above is the ___.
A. unit rise
B. total rise
C. tread width
D. stringer dimension

_____ 5. The recommended riser height for stairs is ___″ to ___″.
    A. 6½; 7
    B. 7; 7½
    C. 7½; 8
    D. none of the above

_____ 6. Tread width plus riser height should be no less than ___″ and no more than ___″.
    A. 15; 16
    B. 16; 17
    C. 17; 18
    D. 16; 18

_____ 7. A ___ square should be used to mark treads and risers on a stringer.
    A. try
    B. framing
    C. combination
    D. bevel

_____ 8. A procedure known as ___ may be used to ensure that all finished riser heights will be the same after treads are added to the stringers.
    A. thickening the treads
    B. decreasing the treads
    C. dropping the stringer
    D. raising the stringer

Identification—Stairway Components

_____ 1. Goose neck

_____ 2. Starting newel post

_____ 3. Open stringer

_____ 4. Riser

_____ 5. Landing newel post

_____ 6. Baluster

_____ 7. Landing

_____ 8. Tread

_____ 9. Handrail

_____ 10. Closed stringer

Unit 64

Stairway Construction

Name _____ Date _____

Completion

_____ 1. ___ is an important concern in stairway design.

_____ 2. The ___ side(s) of any stairway must have handrails.

_____ 3. Minimum space between a handrail and wall is ___".

_____ 4. Stairways in public buildings are usually over ___" wide.

_____ 5. Most residential stairways are ___" to ___" wide.

_____ 6. ___ is the minimum vertical clearance required from any tread on the stairway to any part of the ceiling above the stairway.

_____ 7. Center handrails are placed in public stairways over ___" wide.

_____ 8. Handrails should be at least ___" high.

_____ 9. The width of a stairwell opening should equal the ___ width of the stairs.

_____ 10. Minimum headroom recommended for main stairs is ___.

_____ 11. Handrails should not be over ___" above the tread nosing.

_____ 12. The length of a stairway opening must be determined to provide proper ___.

_____ 13. Stairways with winders are not considered to be as safe as straight-flight stairways because the ___ width varies considerably.

_____ 14. ___ stairways are the most simple type of stairway to build.

_____ 15. Landings must be used to break any stairway rising ___' or more.

_____ 16. Stairways with winders are usually ___-shaped.

_____ 17. Landings should not be over ___' in length when there is no change in stairway direction.

_____ 18. When one side of a prefabricated stairway is open, a(n) ___ stringer is required.

_____ 19. Stairways with ___ stringers are usually prefabricated in a shop.

_____ 20. The minimum headroom recommended for service stairs is ___.

_____ **21.** Recommended riser height for interior stairs is ___″ to ___″.

_____ **22.** The ___ of prefabricated stairways are installed first.

_____ **23.** Exterior stairs with closed treads and risers should have a(n) ___″ slope on the tread for water drainage.

_____ **24.** Treads and risers of prefabricated stairways are set into grooves in the stringers and secured with glue and ___.

_____ **25.** Riser heights for exterior stairs should be ___″ to ___″.

_____ **26.** The printreading abbreviation for lavatory is ___.

Printreading Symbols

Identify this wall symbol.

_____

Unit 65

Post-and-Beam Construction

Name _____ Date _____

Identify this wall symbol.

_____

True-False

T	F	**1.** Three types of post-and-beam construction are residential post-and-beam, timber frame, and beam-frame.
T	F	**2.** Exposed wood beams in post-and-beam construction provide an attractive interior appearance.
T	F	**3.** Metal fasteners are not used in any form of post-and-beam construction.
T	F	**4.** Structural insulated panels should not be used to fill spaces between perimeter wall posts.
T	F	**5.** Exterior walls of post-and-beam buildings may be finished with the same materials used for wood-framed construction.
T	F	**6.** A long post without lateral bracing may buckle.
T	F	**7.** Each end of a transverse roof beam rests on a post.
T	F	**8.** Walls of a post-and-beam building have posts spaced 4′, 6′, or 8′ OC.
T	F	**9.** Plywood floor panels in post-and-beam construction are usually 1⅛″ thick.
T	F	**10.** Timber frame construction uses the same connection methods as post-and-beam construction.
T	F	**11.** Tongue-and-groove planks are normally used to deck a post-and-beam roof.
T	F	**12.** Longitudinal roof beams run the full length of a building.
T	F	**13.** Insulation material may be applied to the top side of a post-and-beam roof.
T	F	**14.** The printreading symbol for light is LIT.

Completion

_____ 1. The basic framework of residential post-and-beam construction consists of ___ posts and ___ beams.

_____ 2. The size of beams used in a post-and-beam floor depends upon the live and dead loads and the ___ between supports.

_____ 3. Post-and-beam construction was once called ___ construction.

_____ 4. Each plank of a post-and-beam floor should span at least ___ opening(s) between floor beams.

_____ 5. The two basic post-and-beam roof designs are ___ and ___.

_____ 6. A(n) ___″ × ___″ post is the minimum size that should be used in a post-and-beam wall.

_____ 7. ___ construction consists of wood roof trusses or rafters connected to vertical posts or timbers.

_____ 8. The main structural members of a post-and-beam roof are the ridge beam, roof ___, and planks or panels used to cover the roof.

_____ 9. ___ partitions of a post-and-beam house are usually erected after the outside walls and roof have been finished.

_____ 10. Modern post-frame structures use square ___ posts as their primary support.

Identification—Timber Frame Framework

_____ 1. Post

_____ 2. Knee brace

_____ 3. Plate

_____ 4. Roof purlin

_____ 5. Collar tie

_____ 6. Diagonal strut

_____ 7. Ridge beam

_____ 8. Wall purlin

_____ 9. Beam

_____ 10. Girt

_____ 11. Strut

Name _____ Date _____

True-False

T   F   **1.** Heavy timber structures have more seismic and wind resistance than most other wood-frame structures.

T   F   **2.** The grain of each lam in a glulam timber runs perpendicular to the length of the timber.

T   F   **3.** Steel connectors are the primary means for tying together the structural components of heavy timber construction.

T   F   **4.** Only tongue-and-groove planks should be used for covering heavy timber roof frames.

T   F   **5.** Timber roof trusses are similar to trusses used in residential construction.

T   F   **6.** Heavy timbers can withstand temperatures that cause steel beams to buckle and twist.

T   F   **7.** Solid lumber is stronger than glulam timbers of equal size.

T   F   **8.** The maximum load stresses on a glulam beam occur at the top and bottom of a beam.

T   F   **9.** Fiber-reinforced glulam timbers cannot be used for all heavy timber operations.

T   F   **10.** Balanced glulam beams should not be used for cantilevered or continuous span applications.

Completion

_____   **1.** As timber burns, the ___ formed on the wood surface helps to protect the unburned wood.

_____   **2.** The ___ lams are placed at the top and bottom of a glulam beam.

_____   **3.** ___ is the movement of a structural component resulting from stress produced by a heavy applied load.

_____   **4.** Steel connectors subjected to water penetration are designed with ___ that allow moisture to drain from the connectors.

_____ 5. The most common glulam purlin width used with preframed panelized systems is ___".

_____ 6. ___ are constructed using a grid system of beams and purlins tied together with steel connecting devices and hubs.

_____ 7. ___, ___, and ___ are most commonly used to manufacture glulam timbers.

_____ 8. Timber roof ___ may be fabricated from solid timbers, parallel strand lumber, or glulam.

_____ 9. For maximum strength, glulam beams are usually placed with the ___ of the lams facing the applied load.

_____ 10. ___ is the slight upward curve in a structural member.

Identification—Glulam Trademark Components

_____ 1. Wood species

_____ 2. Glulam appearance classification

_____ 3. Applicable ANSI standard

_____ 4. Structural use

_____ 5. Mill number

_____ 6. Structural grade designation

_____ 7. Applicable laminating specification

APA—The Engineered Wood Association

Foundation Design for Heavy Construction

Name _____ Date _____

Completion

_____ 1. When soil cannot be excavated, ___ may be driven to support tall buildings.

_____ 2. If building loads and soil conditions do not require a system of piles or caissons, a(n) ___ is usually adequate to support concrete buildings.

_____ 3. Pile-driving equipment has a drop, mechanical, or vibratory ___, which drives the pile or pile casing into the ground.

_____ 4. ___ piles have a long life expectancy under water.

_____ 5. The three major types of piles are ___, ___, and ___.

_____ 6. Shell-less concrete piles can only be used with ___, ___ soil.

_____ 7. Two types of cast-in-place concrete piles are ___ and ___ piles.

_____ 8. ___ caissons have a large bearing area in relation to their shaft diameter.

_____ 9. Bored caissons are usually ___ in shape.

_____ 10. Piles are placed beneath ___ beams, which support bearing walls.

_____ 11. ___ piles are the common type of piles used in concrete heavy construction.

_____ 12. Sheet piles are used primarily to resist ___ pressure.

_____ 13. ___ for shell-type concrete piles remain in place after the concrete is poured.

_____ 14. Tubular steel piles are also known as ___ piles.

_____ 15. Bored caissons may be bored to depths exceeding ___.

_____ 16. Piles may be grouped and joined together with a concrete ___.

_____ 17. ___ piles do not have to penetrate bearing soil.

_____ 18. ___ concrete piles are usually fabricated in a factory.

_____ 19. ___ piles are often used in the construction of wharves and docks.

_____ 20. A(n) ___ is a cylindrical or box-like casing similar to a pile, except it is larger.

True-False

T	F	**1.** Piles are column-like structural members that carry building loads through poor soil to load-bearing soil.
T	F	**2.** Piles may be made of wood, steel, or concrete.
T	F	**3.** Friction piles are the most common type of piles used in heavy construction.
T	F	**4.** Sheet piles are designed to carry vertical loads.
T	F	**5.** Caissons are filled with concrete.
T	F	**6.** Steel piles may be L-shaped or tubular in cross section.
T	F	**7.** The printreading symbol for linoleum is LINO.

Name _____ Date _____

True-False

T	F	**1.** Leaks in forms cause ridges on the surface of the hardened concrete.
T	F	**2.** Cleanout holes in concrete forms are blocked off after the concrete is placed.
T	F	**3.** Forms should be flexible enough to allow movement during the concrete pour.
T	F	**4.** Patented braces are available to brace panel walls.
T	F	**5.** Plyform® panels are not reusable.
T	F	**6.** Single-waler systems do not require studs under the walers.
T	F	**7.** Bucks for door and window openings in concrete walls must be removed before concrete is placed.
T	F	**8.** When constructing basement walls, all inside panel sections are placed first.
T	F	**9.** Braces on wood panel walls are usually placed 6′ to 8′ apart.
T	F	**10.** Slip forms may be used when constructing rectangular buildings.
T	F	**11.** Beams and girders are not placed monolithically.
T	F	**12.** Carpenters place rebar after the outside form panels for concrete walls are set.
T	F	**13.** Beams and girders are not used with the flat-slab system.
T	F	**14.** Gang forms must be lifted by cranes.
T	F	**15.** Tread and riser sizes for concrete and wood stairways may be figured by the same method.
T	F	**16.** Beams support the columns of a concrete building.
T	F	**17.** The printreading abbreviation for living room is LVR.
T	F	**18.** Fiber or steel forms may be used for round columns.
T	F	**19.** A slip form has inner and outer walls.
T	F	**20.** Tall buildings may be poured no more than three stories at a time.

Identification—Outside Wall Forms

_____ **1.** Joist

_____ **2.** Construction joint

_____ **3.** Concrete wall

_____ **4.** Wall form

_____ **5.** Anchor bolt

_____ **6.** Stringer

_____ **7.** Snap ties and walers

_____ **8.** Pour strip

_____ **9.** Shore

_____ **10.** Concrete floor slab

_____ **11.** Walers

_____ **12.** Rebar

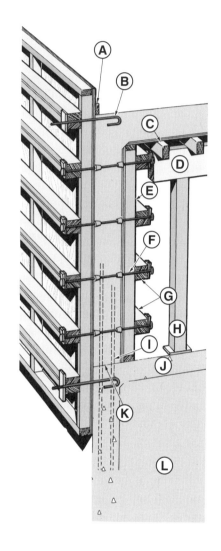

Completion

_____ **1.** Walls, floors, beams, and columns are basic ___ parts of a concrete building.

_____ **2.** ___ fiber forms are frequently used to construct round concrete columns.

_____ **3.** ___ holes are cut at the bottom of the forms to facilitate removal of debris before the concrete is placed.

_____ **4.** ___ is an excellent material to use for concrete forms when curved surfaces are required.

_____ **5.** A wedge-shaped ___ attached to a buck remains in the concrete to serve as a nailing strip for the finish window frame.

_____ **6.** ___ is the most widely used material for building forms.

_____ 7. Form ___ secure and hold opposing form walls together.

_____ 8. A(n) ___ construction joint occurs at each floor level of a tall concrete building.

_____ 9. All floor and roof forms require formwork consisting of a deck supported by ___ and ___.

_____ 10. Setting ___ form walls is also called doubling up the walls.

_____ 11. ___ may be used with a waler system to add strength to the wall or to support the end of a beam.

_____ 12. ___ are placed inside the forms when plans call for long, high concrete walls.

_____ 13. A(n) ___ is a heavy beam that supports other beams.

_____ 14. Concrete may be poured ___ (at one time) for columns, beams, and girders.

_____ 15. ___-clamps may be used to tie column forms in position.

Name _____ Date _____

Completion

_____ **1.** Concrete for heavy construction is delivered to the job site by a(n) ___ truck.

_____ **2.** Climber ___ cranes with buckets are used for pouring operations of large concrete buildings.

_____ **3.** Motorized power ___ may be used where concrete cannot be discharged directly from a bucket.

_____ **4.** Concrete must be ___ to remove excess space between particles.

_____ **5.** ___ strength is the force, in pounds per square inch (psi), a concrete mixture can withstand 28 days after it has been placed.

_____ **6.** ___ is the condition in which the sand-cement ingredients of concrete separate from the gravel.

_____ **7.** A minimum of ___ samples should be taken for a compression strength test on concrete.

_____ **8.** A(n) ___ is used to eliminate high and low areas in a concrete slab.

_____ **9.** ___ is the process of keeping concrete moist long enough to allow hydration to occur.

_____ **10.** Slab-at-grade concrete floors receive direct support from the ___.

Multiple Choice

_____ **1.** A layer of concrete is also known as a ___.
   A. course
   B. lift
   C. form
   D. all of the above

_____ **2.** Internal vibrators are powered by ___.
   A. electricity
   B. gas
   C. compressed air
   D. all of the above

_____ **3.** The slump test measures the ___ of concrete.

    A. strength

    B. consistency

    C. vibratory resistance

    D. none of the above

_____ **4.** The compression strength of concrete is largely dependent upon its ___.

    A. relative drying time

    B. total curing time

    C. water-cement ratio

    D. gravel-sand ratio

_____ **5.** Concrete samples used to determine compression strength are cured for ___ days.

    A. 7

    B. 14

    C. 21

    D. 28

_____ **6.** A concrete floor slab may be ___.

    A. laid directly on the ground

    B. supported by walls and beams

    C. held up by columns

    D. all of the above

_____ **7.** While concrete is being placed, cement finishers strike off the concrete with ___.

    A. bullfloats

    B. jitterbugs

    C. power trowels

    D. a straightedge on screed rails

_____ **8.** An expansion joint in a concrete slab ___.

    A. extends through the slab

    B. provides space for the slab to expand

    C. is placed between the driveway and garage slabs

    D. all of the above

Printreading Symbols

Identify the type of wall shown.

_____

Name _____   Date _____

True-False

T   F   **1.** Lightweight precast members may be manually raised into place.

T   F   **2.** Structural components of prestressed concrete are lighter than similar structural components of normally reinforced concrete.

T   F   **3.** Prefabricated wall panels used in tilt-up construction are usually precast at the plant.

T   F   **4.** Precast T-shaped beams must be prefabricated on the job site.

T   F   **5.** Precast members are usually bolted or welded together.

T   F   **6.** High-tensile steel cables are placed in the forms for prestressed concrete.

T   F   **7.** Prefabricated concrete components may be precast at a plant or on the job site.

T   F   **8.** Tilt-up construction is widely used for buildings with three to six stories.

T   F   **9.** Lift-slab construction combines precast concrete or steel columns with floor slabs cast on the job site.

T   F   **10.** Lifting jacks used in lift-slab construction may be simultaneously operated within a $\frac{1}{16}''$ tolerance.

T   F   **11.** Normal lifting rates of slabs in lift-slab construction are 10′ to 17′ per hour.

T   F   **12.** Columns for lift-slab construction of high-rise structures are erected in sections as the floors are raised.

T   F   **13.** Metal trussed roofs should not be placed on industrial buildings with tilt-up walls.

T   F   **14.** Shear bars inserted through columns and under the slab may be secured by a welded collar.

T   F   **15.** The printreading abbreviation for on center is OC.

Printreading Symbols

Identify the type of wall shown.

_____

Identification—Precast Concrete Structural Members

_____ **1.** Hollow-core slab

_____ **2.** Double-T

_____ **3.** Channel slab

_____ **4.** Single-T

_____ **5.** I-girder

_____ **6.** Columns and pipes

_____ **7.** Monowing

_____ **8.** Inverted T-beam

_____ **9.** Box beam

_____ **10.** Wall and floor panel

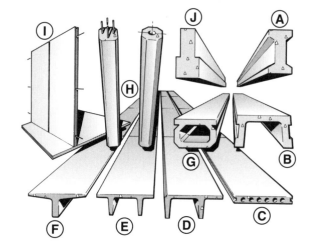

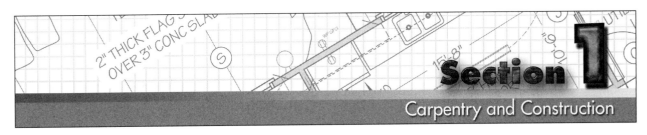

Name _____ Date _____

T   F   **1.** The Associated General Contractors (AGC) represents contractors engaged in heavy construction work.

_____   **2.** A(n) ___ is a modular unit with kitchen and bathroom equipment.

T   F   **3.** Remodeling is considered alteration work.

_____   **4.** An exterior brick wall over a wood stud wall is a(n) ___ wall.

_____   **5.** In post-and-beam construction, the beams are ___.
      A. never exposed
      B. generally exposed
      C. always exposed
      D. for appearance only

T   F   **6.** Heavy construction uses reinforced concrete.

T   F   **7.** Carpenters may not place drywall on walls or ceilings in any part of the U.S.

_____   **8.** Residential construction is a type of ___ construction.
      A. light
      B. heavy
      C. interior
      D. exterior

T   F   **9.** Operating engineers cut, fit, and install glass in windows, doors, skylights, and storefronts.

T   F   **10.** Cement masons may set forms no more than three boards high.

T   F   **11.** Construction craft laborers on a building site may be called on to mix or carry mortar.

_____   **12.** ___ set pumps, turbines, generators, conveyors, and similar types of equipment.

T   F   **13.** High-rise buildings may be referred to as skyscrapers.

_____   **14.** ___ construction is the erection of buildings, railroad trestles, and similar structures.

_____   **15.** ___ steelworkers are employed on high-rise construction to erect the steel framework.

16. How can an individual prepare for the carpentry trade?

_____

_____

_____

_____

_____

_____

_____

17. What points do the United Brotherhood of Carpenters and Joiners of America (UBC) negotiate with contractors' associations?

_____

_____

_____

_____

_____

_____

_____ **18.** In ___ construction, wall units are cast on the floor and raised by a crane.

_____ **19.** A trade association is an organization that represents the ___ of specific products.

T   F   **20.** Modular houses are 95% complete when delivered to the job site.

_____ **21.** Exterior walls of a masonry building may be built of ___.

     A. blocks
     B. bricks
     C. stone
     D. all of the above

_____ **22.** Working in a specific skill area of a trade is known as ___.

_____ **23.** ___ is placed in concrete to provide additional strength.

_____ **24.** The ___ represents contractors in residential and light construction.

_____ **25.** The ___ sections are the basic units of a panel system.

_____ **26.** A ___ is not a prefabricated structural unit.

    A. box beam

    B. glued and laminated beam

    C. prehung door

    D. roof truss

_____ **27.** Which of the following trades finish floors and sidewalks?

    A. plasterers

    B. glaziers

    C. cement masons

    D. construction laborers

_____ **28.** The most common wood-framing method used for most light construction is ___ framing.

    A. post-and-beam

    B. platform

    C. western

    D. none of the above

_____ **29.** In an open panel system, ___ surface(s) of the exterior wall panel are covered with sheathing or insulation board.

    A. inside

    B. outside

    C. top

    D. all of the above

_____ **30.** A modular house is also known as a(n) ___ house.

_____ **31.** Floor slabs are stack-cast around columns and raised by hydraulic jacks in ___ construction.

T   F   **32.** Precast units are poured on the job site.

_____ **33.** ___ work is a type of remodeling in which a change is made to an existing structure.

_____ **34.** Mobile homes are built in accordance with a building code established by the ___.

T   F   **35.** Carpenters today are more specialized than they were in the past.

_____ **36.** The ___ is the national technical trade association of the structural glulam timber industry.

_____ **37.** The base of a mobile home rests on a(n) ___ frame.

    A. steel

    B. plywood

    C. fractional

    D. exposed

38. 3005
      × 101

39.   98
      − 61

40.   638
         711
         461
      + 358

41. 7003
      − 4938

42.   45
      × 12

43. 63,889
      − 50,004

44.   7
           13
          901
           72
      + 1081

45. 45,681
         56,238
      +   57

46.   420
      ×   89

47. 12⟌288

48. 124,578
      −   6584

49. 4⟌456

50. $\begin{array}{r} 8396 \\ +\ 14{,}511 \end{array}$

51. $45\overline{)2025}$

52. $\begin{array}{r} 265 \\ \times\ 381 \end{array}$

53. $\begin{array}{r} 855 \\ 4695 \\ +\ 15{,}874 \end{array}$

54. $438\overline{)141{,}912}$

55. $8\overline{)3456}$

56. $\begin{array}{r} 6284 \\ \times\ 154 \end{array}$

57. $\begin{array}{r} 936 \\ -\ 231 \end{array}$

58. $16\tfrac{5}{8} - 1\tfrac{7}{8} =$

59. $1\tfrac{7}{16} - \tfrac{9}{16} =$

60. $7\tfrac{3}{4} - 2\tfrac{1}{8} =$

61. $10\tfrac{1}{16} - 5\tfrac{11}{16} =$

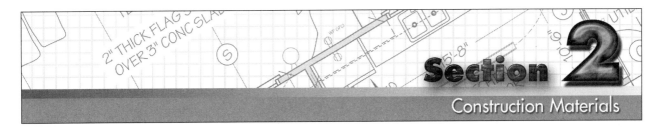
Name _____ Date _____

T   F   **1.** Heartwood lumber is normally weaker than sapwood lumber.

_____ **2.** Dry rot in wood is caused by a ___.
    A. fungus
    B. low humidity level
    C. high humidity level
    D. lack of water

_____ **3.** Veneers are graded according to their ___.
    A. natural growth characteristics
    B. size and number of repairs during manufacture
    C. appearance
    D. all of the above

_____ **4.** ___ is composed of wood laminations bonded with adhesive.

T   F   **5.** Trees have one bark layer.

T   F   **6.** Sapwood is darker than heartwood.

_____ **7.** A board measuring $1'' \times 4'' \times 6'$ contains ___ BF.

_____ **8.** The most common wood defect in a board is a(n) ___.

_____ **9.** Saw sized lumber is abbreviated as ___.

_____ **10.** A widely used structural wood panel is ___.

T   F   **11.** Most lumber is produced by quartersawing.

_____ **12.** Softwood lumber is normally ___ dried.

T   F   **13.** The annual rings of a tree trunk are farther apart during dry seasons.

_____ **14.** Wood does not decay when its moisture content is below ___%.

T   F   **15.** The new cell growth of a tree is formed in the cambium.

_____ **16.** The central core of a tree is the ___.

T   F   **17.** Deciduous trees normally lose their leaves in the fall.

T   F   **18.** Southern pine is the predominant rough construction wood used in the southeastern part of the United States.

_____ **19.** Lumber measurements are stated in the following order: ___.
         A. thickness, width, length
         B. thickness, length, width
         C. width, thickness, length
         D. length, thickness, width

T   F   **20.** Medium density fiberboard is a nonstructural panel product.

_____ **21.** ___ are evergreen trees having needles and cones.

T   F   **22.** Only hardwood lumber is used for rough construction.

_____ **23.** Timbers are no smaller than ___″ wide by ___″ thick.

T   F   **24.** Most hardwood trees grow in the eastern part of the United States.

_____ **25.** How much will 118′ of cove molding at $.36 per lineal foot cost?

T   F   **26.** The most commonly used adhesives in the construction industry are glues and mastics.

_____ **27.** Heads on wood screws are flat, ___, or oval.

T   F   **28.** Toggle bolts have machine screw threads.

T   F   **29.** Double-head nails are used primarily for permanent construction.

T   F   **30.** Stove bolts have a square shoulder below their heads.

_____ **31.** Toggle bolts have ___ screw threads.

_____ **32.** Standard sizes of plywood panels are 4′ × 8′, 4′ × 10′, and ___′ × ___′.

T   F   **33.** Lignin is an artificial wood-binding agent used for adhesive.

T   F   **34.** Studs are 2″ × 4″ or 2″ × 6″ pieces of lumber that are 12′ or shorter.

T   F   **35.** Yellow poplar is a hardwood.

_____ **36.** The actual size of a 2″ × 4″ × 8′ stud is ___.
         A. 1¼″ × 3¼″ × 8′
         B. 1½″ × 3½″ × 8′
         C. 1⅝″ × 3⅝ ″ × 8′
         D. 2″ × 4″ × 8′

T   F   **37.** Nails have greater holding power than screws.

_____ **38.** A contractor places an order for the following:
         132 pieces of ¼″ luan plywood at $16.49 each
         500 studs at $1.88 each
    What is the total material cost?

T   F   **39.** Plywood always has an even number of layers.

T   F   **40.** Interior panels may be used only for interior application.

T   F   **41.** Three grades of hardboard are premium, standard, and common.

T   F   **42.** Common nails are most often used for wood-frame construction.

_____**43.** The abbreviation for lumber rated as firsts and seconds is ___.

_____**44.** One-fourth of 16′-8″ is ___.

_____**45.** The watery fluid found in a freshly cut tree is ___.

_____**46.** A board's deviation from a flat plane, edge to edge, is a(n) ___.

---

47. $16'\text{-}8'' + 32'\text{-}1'' =$

48. $\frac{3}{8} \times 3 =$

49.
$$7'\text{-}10\ ''$$
$$3'\text{-}5\frac{1}{2}''$$
$$+\ 2'\text{-}7\ ''$$

50. $1\frac{7}{16} + 9\frac{5}{8} + 1\frac{3}{4} =$

51. $3\frac{1}{2} + 2\frac{5}{8} + 9\frac{5}{16} =$

52. $13.82 + 6.91 + 7.85 =$

53. $.055 + 7.61 + 10.094 =$

54. $23.83 + 81.17 + .05 =$

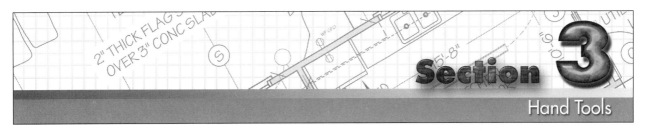
Name _____ Date _____

_____ 1. A(n) ___ square is used for marking 45° and 90° angles.

T   F   2. Most accidents involving hammers and hatchets result in injuries to the user's fingers.

_____ 3. ___-claw hammers are used primarily for finish work.

T   F   4. Most handsaws have a straight back.

T   F   5. Crosscut saws should be held at a 60° angle to the work.

_____ 6. The ___ of the saw teeth helps prevent binding of the saw blade.

T   F   7. Compass saws are used to cut curves and irregularly shaped lines.

T   F   8. Ratchet braces can be operated by turning them only in a clockwise direction.

T   F   9. Jack planes may be used for general-purpose planing.

T   F   10. Rasps have chisel-shaped teeth.

T   F   11. Block planes are used for planing small, narrow surfaces.

T   F   12. Cabinet scrapers are used before sandpapering for a final finish.

_____ 13. The cutting edge of a plane iron for a bench plane is usually ground at a ___° to ___° angle.

T   F   14. All hand planes are equipped with handles for easy grasping.

T   F   15. A dovetail saw is similar to a backsaw, but it is smaller.

_____ 16. A(n) ___ is used to snap lines on flat surfaces.

_____ 17. Nails have more holding power when driven ___ the grain.

_____ 18. The size of a screwdriver is determined by the ___ of its ___.

_____ 19. The shoulder of the pilot hole should be countersunk when recessing ___ screws.

_____ 20. Pressure should always be applied against the ___ jaw of an adjustable wrench.

T   F   21. An 11-point saw blade has larger teeth than an 8-point saw blade.

_____ 22. Crosscut saws used for finish work normally have ___ to ___ points per inch.

_____ **23.** The number printed on the blade of a handsaw indicates the ___.
    A. maximum depth of cut
    B. minimum width of cut
    C. blade length
    D. number of teeth per inch

_____ **24.** The number on the tang of an auger bit indicates ___.
    A. maximum depth of cut in eighths of an inch
    B. maximum depth of cut in sixteenths of an inch
    C. size in sixteenths of an inch
    D. size in eighths of an inch

_____ **25.** ___ tools are used to hold and support materials.

T   F   **26.** End grain of a piece of wood should be planed from each edge to prevent splitting.

T   F   **27.** Double edge and chamfer rabbet planes are used for finish smoothing operations.

T   F   **28.** Serrated blades of forming tools can be easily sharpened.

_____ **29.** Truing the edges of boards before they are fitted together is known as ___.
    A. jointing
    B. smoothing
    C. layering
    D. edging

_____ **30.** How many sixteenths are in ¾?

T   F   **31.** Jointer planes are normally 24″ to 28″ in length.

_____ **32.** A(n) ___ is used to produce a burr on a scraper blade.

_____ **33.** A(n) ___-hatchet is used when building wood forms for concrete.

T   F   **34.** Flooring chisels are used primarily for rough work.

_____ **35.** A set of ___ is used to lay out circles of any size.

T   F   **36.** The term _plumb_ as used in carpentry refers to establishing a vertical plane.

_____ **37.** A(n) ___ should be used to sink nail heads below the surface.

T   F   **38.** The size of a hammer is determined by its total weight.

_____ **39.** A(n) ___ hammer can be used to drive nails flush without marring the surface.

T   F   **40.** Screwdrivers with long blades allow greater force to be applied.

_____ **41.** The three basic types of screwdrivers used in the construction trades are ___, ___, and ___.

T   F   **42.** Wrenches tend to slip more often when a pulling pressure is applied.

T　　F　　**43.** A carpenter's level may be used in a horizontal or vertical position.

T　　F　　**44.** A combination square may be used to mark a 45°, 60°, or 90° angle.

T　　F　　**45.** Inches and sixteenths of an inch are marked along the outside edges of the back side of the blade and tongue of a framing square.

_____ **46.** A(n) ___ hatchet is used for installing roof shingles.

T　　F　　**47.** A scriber has two legs.

T　　F　　**48.** The neck of a hammer is the smallest portion of the handle.

_____ **49.** When toenailing boards, the nail should be driven so that approximately ___% of the nail is in each board.

_____ **50.** A(n) ___ hatchet has a curved nailing face that will dimple the wallboard without breaking the paper covering.

T　　F　　**51.** Drilling pilot holes may be necessary when driving wood screws into harder woods.

---

52. $3\frac{1}{2} - 2\frac{3}{4} =$

53. $14\frac{7}{8}'' - 3\frac{1}{4}'' =$

54. $2\frac{1}{2} \times 2\frac{1}{2} =$

55. $\begin{array}{r} 14' - 8'' \\ - \ \ 6' - 10'' \\ \hline \end{array}$

56. $\begin{array}{r} 6' - 8'' \\ 3' - 4'' \\ + \ \ 12' - 2'' \\ \hline \end{array}$

57. $48 \div \frac{3}{4} =$

58. $25.55 \times 30.10 =$

59. $6' - 8\frac{1}{2}'' + 1' - 6\frac{1}{4}'' =$

60.
$$
\begin{array}{r}
387 \\
810 \\
911 \\
+\ \ 432 \\
\hline
\end{array}
$$

61.
$$
\begin{array}{r}
150.75 \\
75.70 \\
120.55 \\
+\ \ 550.10 \\
\hline
\end{array}
$$

62.
$$
\begin{array}{r}
8' - 10'' \\
-\ \ 3' - 11'' \\
\hline
\end{array}
$$

63. $\dfrac{3 \times 16 \times 10}{4}$

64.
$$
\begin{array}{r}
16' - 10\frac{1}{4}'' \\
-\ \ 9' - 7\frac{3}{4}'' \\
\hline
\end{array}
$$

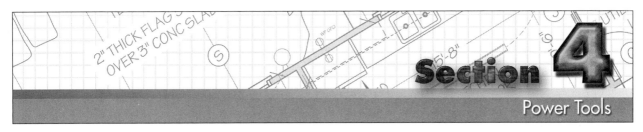
Name _____ Date _____

T   F   **1.** Electric tools should be disconnected when not in use.

T   F   **2.** A water pipe provides an excellent ground.

_____ **3.** Circular electric handsaws are adjustable to cut angles from ___° to ___°.

_____ **4.** A radial arm saw is also known as a(n) ___ saw.

_____ **5.** Electric tools having a conductor cord with a three-prong plug must be ___.

_____ **6.** A frame-and-trim saw cannot make a ___.
        A. rip
        B. crosscut
        C. miter
        D. bevel

T   F   **7.** Faster drill speeds should be used for harder materials.

T   F   **8.** Spade bits are also known as broad bits.

T   F   **9.** Carpenters use routers primarily for mortising and shaping operations.

_____ **10.** A ___ plane should be used to smooth end grain.
        A. face
        B. block
        C. jack
        D. jointer

T   F   **11.** Penetration depths of powder-actuated tools may be controlled by the load of cartridge used.

_____ **12.** Powder loads are ___ for ease of identification.

T   F   **13.** Threads of nuts and couplings on oxygen lines are always right-handed.

T   F   **14.** Carpenters may weld straps and hangers in place.

_____ **15.** ___ arc welding is the type of welding most commonly used by carpenters.

_____ **16.** Ultraviolet and infrared rays from welding can cause damage to the ___.

_____ **17.** In ___ welding, metal is fused together by the heat of a welding flame.

_____ **18.** Electric arc welding machines may be ___.
 A. AC
 B. DC
 C. AC-DC
 D. all of the above

_____ **19.** In the interest of safety, a table saw blade should be no more than ___″ to ___″ above the material being cut.

_____ **20.** The following pieces of ceiling molding are required for a den:
 two – 16′-4″
 two – 12′-2″
 What is the total number of lineal feet of ceiling molding required?

T   F   **21.** Hole saws may be used to drill holes in glass.

_____ **22.** ___ should be worn when welding metals that may give off toxic fumes.

_____ **23.** An acetylene hose is always colored ___.

_____ **24.** The greatest hazard related to electric tools is ___.

T   F   **25.** Cuts on long pieces should be made between sawhorses.

_____ **26.** The blade of a reciprocating saw has a(n) ___ motion.

_____ **27.** A(n) ___ saw is used by carpenters to trim out buildings.

T   F   **28.** Hammer drills cannot be used to drill steel.

T   F   **29.** Nails for pneumatic nailers may be packaged in coils or strips.

_____ **30.** ___ trimmers are used to trim the edges of countertops.

T   F   **31.** The exposed bit represents the most potential danger when using a router.

_____ **32.** Concrete thickness should be at least ___ times the fastener shank penetration.

T   F   **33.** Saber saw blades cut on the downstroke.

_____ **34.** ___ blades should be used for cutting metals.
 A. Combination
 B. Ripping
 C. Crosscutting
 D. none of the above

T   F   **35.** Abrasive blades should be used for cutting masonry materials.

_____ **36.** The recommended size of radial arm saws for construction ranges from ___″ to ___″.
 A. 10; 12
 B. 12; 14
 C. 14; 16
 D. 16; 18

_____ **37.** How many 1¾″ strips may be cut from a 10″ wide board? (Allow ⅛″ for each saw kerf.)

_____ **38.** ___ may be used in conjunction with dado blades to increase the width of the dado cut.

_____ **39.** The size of a radial arm saw is determined by the ___.
   A. horsepower rating of the saw
   B. length of the overarm
   C. largest blade it will accommodate
   D. size of the table

_____ **40.** A frame-and-trim saw can cut stock up to ___″ thick and ___″ wide.
   A. 1; 6
   B. 2; 6
   C. 1; 12
   D. 2; 12

   T   F   **41.** Feeler bits can drill holes up to 24″ deep.

_____ **42.** How many blocks 22½″ long can be cut from a 2″ × 4″ × 8′ stud?

_____ **43.** ___, including proper eye, hearing, and respiratory protection, should be worn during sanding operations.

_____ **44.** A(n) ___ may be used to cut curved lines and circular or rectangular openings.

_____ **45.** The air pressure required to operate a pneumatic tool is adjusted by a(n) ___.
   A. air gauge
   B. restrictive hose fitting
   C. regulator
   D. none of the above

_____ **46.** ___ can usually be prevented on a radial arm saw by tilting the saw table slightly back when the unit is set up.
   A. Crawl
   B. Ripping
   C. Freehand cutting
   D. none of the above

_____ **47.** Compressor air hoses more than ___″ in diameter must have a safety device to stop air flow in case of a hose failure.

_____ **48.** Portable power drills designed for heavy-duty construction have a(n) ___ handle.
   A. spade
   B. inverted
   C. adjustable
   D. none of the above

_____ **49.** The following molding is required for a bedroom:
    52′ of baseboard @ $.47 per lineal foot
    52′ of shoe mold @ $.14 per lineal foot
  What is the total cost of the molding?

_____ **50.** Slash grain in wood refers to a ___.
  A. straight, evenly spaced grain
  B. straight, unevenly spaced grain
  C. board having no particular grain pattern
  D. change in grain direction

_____ **51.** ___ saws are used to cut heavy timber and pilings.
  A. Table
  B. Circular
  C. Bayonet
  D. Chain

_____ **52.** A(n) ___ sander has a circular and oscillating motion.
  A. belt
  B. orbital
  C. random orbital
  D. detail

---

53. $1\frac{3}{4}$ of 20 =

54. .375 × .50 =

55. 32.5
  × 10.5

56. 25)‾5.050‾

57. $\frac{7}{8} \times 1 \times \frac{3}{4} =$

58. 215 × .540 =

59. $\frac{3}{8} \times 2 =$

60. 16′ - 8½″
  − 8′ - 9 ″

Name _____ Date _____

T   F   **1.** Scaffold planks must extend at least 6″ past the end supports.

T   F   **2.** As a general rule, scaffolds should be able to support eight times the maximum load to which they will be exposed.

_____   **3.** A ___ is used for tying off members to be raised.
       A. half-hitch
       B. bowline
       C. timber
       D. none of the above

T   F   **4.** Wet ladders do not conduct electrical current.

_____   **5.** A(n) ___ is normally used to strip rocks and topsoil for the excavation site.
       A. motor grader
       B. bulldozer
       C. excavator
       D. none of the above

_____   **6.** ___ are dismantled into sections before they are lowered to the ground.
       A. Forklifts
       B. Climbing cranes
       C. Free-standing cranes
       D. Earth augers

T   F   **7.** The most common injury suffered by construction workers is strain or over-exertion.

T   F   **8.** A wood-burning fire is a Class A fire.

T   F   **9.** Slips and falls from elevated work surfaces and ladders account for one-third of all construction injuries.

T   F   **10.** Work surfaces 6′ or higher should be guarded by railings.

T   F   **11.** Water can be used to extinguish Class A fires.

_____   **12.** Suspension scaffolds are supported by wire ropes that must support at least ___ times the maximum intended load.
       A. four
       B. six
       C. eight
       D. ten

T   F   **13.** Toeboards are installed to protect workers' toes.

T   F   **14.** A metal scaffold may be erected within 8′ of noninsulated power lines.

_____ **15.** Safety nets must be used for work ___′ or more above the ground when a worker is not otherwise protected.
     A. 8
     B. 12
     C. 15
     D. 25

T   F   **16.** Graders are normally used for final grading on large construction sites.

_____ **17.** The banks of an excavation 20′ deep in average soils should slope ___°.
     A. 10
     B. 30
     C. 45
     D. 60

_____ **18.** ___ within 5′ of one another may be used for shoring in hard, compact soil.

T   F   **19.** Guardrails should be nailed across all wall openings from which there is a drop exceeding 4′.

T   F   **20.** The noise level of certain operations on construction jobs can cause permanent ear damage over a period of time.

T   F   **21.** The amount of excavation on a job site is independent of the size of the structure to be erected.

T   F   **22.** Guardrails are not required around floor openings that are framed for stairwells.

T   F   **23.** Workers may ride on a load only if it is attached to the cable of a crane.

_____ **24.** Lumber piles that are to be handled manually should not exceed ___′ in height.
     A. 4
     B. 8
     C. 10
     D. 16

_____ **25.** Backhoe loaders are equipped with ___ to provide stability during trenching operations.

_____ **26.** Two carpenters require 1¾ hr each to construct a small, single-pole wood scaffold. The carpenters are paid $17.50/hr. What is the labor cost of the scaffold?

T   F   **27.** OSHA lists safety procedures designed to protect workers on the job.

T   F   **28.** Respiratory protection is required when workers are exposed to airborne hazards.

T   F   **29.** A square knot can be used to fasten together the ends of two ropes of the same thickness.

_____ **30.** A runway for motor-driven concrete buggies is constructed 8′-0″ wide and 42′-0″ long. What is the total area of the runway?

T   F   **31.** Class B fires occur with flammable liquids.

T   F   **32.** Class D fires occur with live electrical equipment.

T   F   **33.** Lumber with loose knots may be used for scaffolds.

_____ **34.** Tie-ins and ___ provide support for a scaffold and prevent it from tipping.

_____ **35.** Sheet piling for trenches over 8′ deep must be at least ___″ thick.

_____ **36.** Platform planks for scaffolds should overlap a minimum of ___″.

T   F   **37.** Hard hats and fall-arrest equipment are optional when working on elevated work platforms or buckets of aerial lifts.

T   F   **38.** An excavator may be used for loading operations.

_____ **39.** ___ scaffolds are the primary type of scaffold used in construction.

_____ **40.** A bulldozer and dump trucks are required to strip and remove topsoil from a building site. The contractor has two subcontractors who have made bids on the job. Subcontractor A will do the job in 14 hours at $132.00/hr and subcontractor B will do the job in 12 hours at $148/hr. Which subcontractor has the lower bid?

_____ **41.** An equipment company offers delivery, pickup, and rental of scaffolding required for a job at $250.00 per week. Scaffold erection and dismantling at $18.35/hr is extra and would require four hours for erection and two hours for dismantling. What is the total cost of rental, delivery, erection, dismantling, and pickup of the scaffolding for one week?

T   F   **42.** A scaffold is a permanent work platform.

_____ **43.** Commercially available ladders are made of ___.
   A. wood
   B. fiberglass
   C. aluminum
   D. all of the above

T   F   **44.** Carpenters may set lines and lay out areas for excavation and trenching.

_____ **45.** What does the following hand signal mean?
   A. raise
   B. lower
   C. swing
   D. stop

T   F   **46.** Hand signals may be given by any worker on a job site.

T   F   **47.** Interlocking sheet piling may be reused.

T   F   **48.** Class A and Class B fires should be extinguished with water.

_____ **49.** When lifting heavy objects, a worker should bend at the ___.

T   F   **50.** Flammable materials should not be stored on the immediate job site.

_____ **51.** What does the following hand signal mean?
- A. hoist
- B. raise the boom
- C. raise the boom and lower the load
- D. extend the boom

T   F   **52.** Wooden or metal brackets may be used for carpenters' scaffolds.

_____ **53.** The distance from the work platform to the upper surface of the top rail should be ___″ to ___″.
- A. 32; 48
- B. 36; 42
- C. 36; 48
- D. 38; 45

T   F   **54.** An inside-climbing tower crane may be set up in an elevator shaft.

_____ **55.** High fences known as ___ are often erected around a job site to prevent entry by unauthorized personnel.

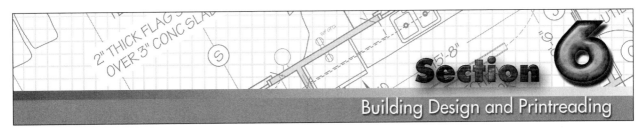
Name _____ Date _____

_____ **1.** The four basic shapes of one-family dwellings are ___, ___, ___, and ___.

_____ **2.** Which of the following is not given on a plot plan?
  A. elevations
  B. swales
  C. driveway locations
  D. number of windows

T   F   **3.** Size and spacing of floor joists are given on the foundation plan.

T   F   **4.** Foundation footings are visible on the foundation plan.

T   F   **5.** Floor plans contain lines showing switch control of lights.

T   F   **6.** Material finish for outside surfaces of walls and roofs is given on elevation drawings.

_____ **7.** The unit rise of a roof is found on the ___ drawings.

_____ **8.** A(n) ___ view of a foundation shows the shape of the foundation wall.

_____ **9.** ___ for base cabinets are 36″ above the finished floor.

_____ **10.** A building lot for a one-family dwelling is 125′ wide and 190′ deep. How many square feet does the lot contain?

_____ **11.** ___ are legal documents that help clarify working drawings.

T   F   **12.** A habitable room is a room used for living purposes.

_____ **13.** The abbreviation GA on a set of prints denotes ___.
  A. gauge
  B. garage
  C. grade access
  D. general area

T   F   **14.** Prints give a pictorial view of each part of a building.

T   F   **15.** A centerline on a print is indicated by a series of short dashes.

_____ **16.** The symbol Ⓙ on a print indicates a(n) ___.

_____ **17.** Easements on plot plans may indicate provisions for ___.
      A. planting trees
      B. use of public utility companies
      C. front setbacks
      D. all of the above

_____ **18.** Finish grades on a plot plan are based upon their relation to the ___.
      A. existing streets
      B. property lines
      C. benchmark
      D. roof height

_____ **19.** A(n) ___ refers to the grade for channeling water away from a building.

_____ **20.** A(n) ___ plan shows the finish grade at all corners of a lot.

_____ **21.** Commercial zoning regulations prohibit a building from occupying over 45% of a building lot. What is the maximum size one-story building that may be built on a rectangular lot measuring 120′ wide and 184′ deep?

_____ **22.** The three major types of zones in larger communities are ___, ___, and ___.

T   F   **23.** A scale of ⅛″ = 1′-0″ is the most commonly used scale on a set of architectural prints.

T   F   **24.** Architectural plan symbols may represent materials, fixtures, or structural parts of a building.

T   F   **25.** Dimension lines should be terminated by arrowheads or dots.

_____ **26.** A Ⓣ on a set of plans denotes a(n) ___.

_____ **27.** Different sides of a building shown on a set of plans may be referred to by ___ directions.

_____ **28.** ___ lines on a plot plan may show the existing and/or finish grade.

T   F   **29.** Benchmarks must be located on the property shown on a plot plan.

_____ **30.** The grade figure on a benchmark is established as 100.00′. The finish floor, which is 48′-0″ away, is given as 106.4′. The finish floor is ___′ above the benchmark.

T   F   **31.** Widths of sidewalks and driveways may be found on the plot plan.

T   F   **32.** Floor plans show the finish for interior walls.

T   F   **33.** The location of kitchen cabinets is given on the floor plan.

T   F   **34.** The length of exterior walls on a floor plan is shown by dimension lines.

T   F   **35.** Arrows are used on a floor plan to show the direction in which ceiling joists run.

_____ **36.** Foundation footings on elevation drawings are shown with ___ lines.

_____ **37.** Wire mesh in a slab is shown by ___ lines.

T   F   **38.** Section views may be taken with vertical or horizontal cuts.

T   F   **39.** Section views may be taken on floor plans and foundation plans.

_____ **40.** A buyer agrees to pay $1.68/sq ft for a 124′ × 182′ lot. What is the cost of the lot?

T   F   **41.** The benchmark for a lot should be located at the lowest point on the lot.

_____ **42.** A house that is 28′ deep is set back 40′ from the property line of a lot, which is 150′ deep. How deep is the backyard?

T   F   **43.** Written notes on architectural plans are abbreviated whenever possible.

_____ **44.** S_3 on a floor plan indicates a(n) ___.

_____ **45.** A set of plans is drawn at a scale of ¼″ = 1′-0″. The South wall is drawn 7½″ long. What is the length of the South wall?

_____ **46.** Benchmarks are established by a ___.
    A. surveyor
    B. contractor
    C. carpenter
    D. none of the above

_____ **47.** A benchmark is also known as a job ___.

_____ **48.** The distance from the property line to the front of a building is known as the ___.

_____ **49.** A light manufacturing company pays $37,039.20 for a commercial lot that measures 122′ wide and 184′ deep. What is the land cost per square foot?

T   F   **50.** A plot plan usually includes an arrow designating North.

_____ **51.** Most buildings are either traditional or ___ in appearance.

_____ **52.** Prints show ___ views of each part of a building.
    A. pictorial
    B. isometric
    C. oblique
    D. orthographic

_____ **53.** Measurements on drawings to show distances between different points are known as ___.

_____ **54.** The symbol (SD) on a print indicates a ___.
- A. switching device
- B. swinging door
- C. single doorbell
- D. none of the above

_____ **55.** The abbreviation REF on a print stands for ___.
- A. roofing
- B. a refrigerator
- C. reinforcement
- D. a rough opening

_____ **56.** A building lot is 210′ along the North and South sides and 265′ along the East and West sides. This lot is ___.
- A. square
- B. rectangular
- C. L-shaped
- D. none of the above

_____ **57.** Ten 2″ × 10″ × 24′-0″ joists are required to support the floor of a living-dining room combination. At $2.19 per lineal foot, what is the cost for these joists?

_____ **58.** The abbreviation BT on a set of prints is seen only in the ___.
- A. bedroom
- B. bathroom
- C. kitchen
- D. garage

_____ **59.** A(n) ___ line is used on a set of prints to indicate the shortened view of a part that has a uniform shape.

_____ **60.** A set of prints shows that a 24′-0″ × 36′-0″ area must be excavated to a depth of 8′-0″ for a basement. How many cubic yards of earth will be removed?

Name _____ Date _____

T   F   **1.** Builder's levels have telescopes ranging from 12 power to 32 power.

T   F   **2.** An engineer's leveling rod is graduated in feet only.

T   F   **3.** Hours and minutes are used to express fractions of a degree.

_____   **4.** A builder's level has crosshairs inside the ___ of the telescope.

_____   **5.** A(n) ___ leveling rod is graduated in feet, inches, and eighths of an inch.

_____   **6.** ___ leveling screws are used to make a rough adjustment for an automatic level.
            A. Two
            B. Three
            C. Four
            D. none of the above

T   F   **7.** The right angle is the most frequently used angle in construction work.

_____   **8.** Both arms straight up above the head indicate that the rod ___.
            A. is plumb
            B. should be raised
            C. is too high
            D. none of the above

T   F   **9.** Strong air disturbances have no effect on the accuracy of a laser light.

_____   **10.** ___ instruments combine survey technology with digital data processing.

T   F   **11.** Electronic distance measurement has an accuracy of 0.1′ without the use of a measuring tape.

_____   **12.** A architect's leveling rod is graduated in ___, ___, and ___.
            A. feet; tenths of a foot; hundredths of a foot
            B. feet; inches; eighths of an inch
            C. meters; centimeters; millimeters
            D. none of the above

T   F   **13.** A manual laser level is leveled in a manner similar to traditional surveying equipment.

T   F   **14.** The target for a laser transit-level is known as a receiver.

_____ **15.** Ninety degrees is ___ of 360°.
    A. ¼
    B. ½
    C. ¾
    D. none of the above

    T   F   **16.** The horizontal circle of a transit-level cannot be moved by hand.

_____ **17.** Minutes and seconds are used to express fractions of a(n) ___.

    T   F   **18.** Centering the bubble is the last step when leveling a builder's level.

_____ **19.** A benchmark is also known as a(n) ___.

_____ **20.** A ___ is used when a builder's level must be set up over a specific point.
    A. scale
    B. level
    C. plumb bob
    D. none of the above

_____ **21.** A horizontal ___ screw allows the telescope of a builder's level to be moved slightly to the left or right.

_____ **22.** ___ foot numbers are the largest numbers on a leveling rod.

_____ **23.** A horizontal ___ screw is used to hold the builder's level in a fixed horizontal position.

    T   F   **24.** Figures and graduations between foot numbers on a leveling rod are usually printed in the color white.

_____ **25.** A compensated self-leveling laser level has fully automatic leveling operations.

_____ **26.** Both arms extended to the sides indicates that the target is ___.
    A. to the left
    B. to the right
    C. on grade
    D. none of the above

    T   F   **27.** Both arms extended over the head indicates that the target is plumb.

    T   F   **28.** The beam of a laser level is approximately 1⅜″ in diameter.

    T   F   **29.** Class II lasers that are commonly used on construction sites can sometimes present a health hazard.

_____ **30.** A tool commonly used to check and establish grades and elevations and to set level points is a(n) ___.

_____ **31.** The intersecting scales of a transit-level used to measure horizontal angles are the ___ and ___ scales.

    T   F   **32.** A laser level is accurate to within 1⁄16″ at a range of 100′.

T   F   **33.** The telescope of a transit-level can be moved only in a horizontal direction.

________________________ **34.** The vernier scale of a transit-level has ___ graduations of 0 to 60 minutes at each side of the zero index.
    A. three
    B. four
    C. six
    D. twelve

________________________ **35.** Transit-levels have a vertical arc for measuring ___ angles.

________________________ **36.** Transit-levels have a(n) ___ or clamp to hold the telescope in a fixed position.

T   F   **37.** Leveling rods are made of wood, plastic, or sheet metal.

________________________ **38.** The three-legged support on which a builder's level is mounted is known as a(n) ___.

T   F   **39.** A laser level can be used to plumb horizontal items.

T   F   **40.** Laser transit-levels cannot be wall-mounted.

________________________ **41.** The telescope of a builder's level is normally adjusted with ___ leveling screws while checking the spirit level.
    A. two
    B. four
    C. six
    D. eight

T   F   **42.** Only one person is required to perform layout operations with a laser transit-level.

T   F   **43.** The legs of a builder's level may be extended if required.

T   F   **44.** A laser transit-level cannot be used for leveling over long distances.

________________________ **45.** A(n) ___ is the vertical measuring device held by a second person when a builder's level is used to check or establish grades and elevations.

---

46. $112°\text{-}20' + 15'\text{-}10'' =$

48. $180° + 70° + 10° + 32° =$

47. $\begin{aligned} &19°\text{-}22'\text{-}36'' \\ + \ &31°\text{-}16'\text{-}52'' \end{aligned}$

49. $58'\text{-}40'' - 16'\text{-}31'' =$

50. $120° + 17°\text{-}15' =$

53. $1°\text{-}22'\text{-}17''$
    $+ \quad 39'\text{-}47''$
    $\overline{}$

51. $36°\text{-}15' - 14°\text{-}12'\text{-}45'' =$

54. $90°\text{-}30'\text{-}10''$
    $- \quad 5°\text{-}20'\text{-}40''$
    $\overline{}$

52. $120°\text{-}40'\text{-}30'' - 90°\text{-}55' =$

55. $120°\text{-}15'\text{-}30'' + 18°\text{-}20'\text{-}10'' =$

Name _____ Date _____

_____ **1.** Most buildings are constructed on soils that are classified as ___, ___, ___, or ___.

_____ **2.** Local building ___ normally specify the depth of foundation footings.

_____ **3.** The moisture content of plywood used in wood foundations should not exceed ___%.

T   F   **4.** Areaways must project below the finish-grade and above the bottom of the window.

T   F   **5.** Concrete is the strongest and most durable material used for foundations.

_____ **6.** Three cubic yards of concrete contain ___ cu ft.

_____ **7.** Rectangular or battered forms can be built by the ___ method.
    A. built-in-place or panel
    B. monolithic
    C. trench
    D. all of the above

T   F   **8.** The minimum recommended thickness for sidewalks is 6″.

T   F   **9.** Driveways should have a minimum cross slope of ½″ per foot for drainage.

T   F   **10.** Basement foundation walls should be waterproofed from the edge of the footing to the finish-grade line.

T   F   **11.** Galvanized iron may be used for termite shields.

_____ **12.** ___ are normally required in concrete or masonry foundation walls constructed in seismic risk zones.

_____ **13.** A form with an area of ___ sq ft can be poured 12″ deep with one cubic yard of concrete.

T   F   **14.** The bent end of an anchor bolt should be embedded at least 7″ into reinforced concrete.

_____ **15.** How many yards of concrete are required to pour a 6″ slab 6′ wide and 9′ long?

T   F   **16.** Concrete forms are temporary structures.

T   F   **17.** Frames constructed on-the-job for door and window openings in concrete walls are known as bucks.

_____ **18.** ___ can be snapped off at breakback points after the forms are stripped.

T   F   **19.** Walers, studs, and braces are usually cut from 2 × 6s.

_____ **20.** A shallow ___ should be cut in a batterboard to prevent the lines from moving.

_____ **21.** The footings and walls of a T-foundation having low walls and a crawl space are poured ___.
    A. as one unit
    B. footings first
    C. walls first
    D. in no particular order

_____ **22.** Key strips used to form keyways in footings are placed in forms ___.
    A. before the pour and are not removed after the concrete hardens
    B. before the pour and are removed after the concrete hardens
    C. during the pour and are not removed after the concrete hardens
    D. during the pour and are removed after the concrete hardens

_____ **23.** Piers that serve as a base for steel columns may be ___.
    A. round
    B. square
    C. battered
    D. all of the above

_____ **24.** A piece of wood used to indicate the proper level of a concrete pour is known as a(n) ___.

T   F   **25.** Pressure-treated lumber or redwood is recommended for sill plates because of its decay and insect resistance.

T   F   **26.** Stepped foundations cannot be used with a full basement.

T   F   **27.** Screeding operations may be performed manually.

_____ **28.** ___ joints provide spacing between dissimilar construction.

T   F   **29.** Forms for outdoor slabs are normally constructed of 2 × 4s or 2 × 6s.

T   F   **30.** Double-car driveways are normally 15′ to 18′ wide.

T   F   **31.** When 2″ thick planks are used for sheathing, walers are not required.

T   F   **32.** All plastic drainpipe is perforated to allow seepage.

_____ **33.** ___ consume the interior portions of wooden boards.

T   F   **34.** Clay particles compress more than sand particles when subjected to heavy pressure.

T   F   **35.** Concrete has more compression strength than lateral strength.

_____ **36.** A(n) ___ is the base for a wall.

_____ **37.** The depth to which trenches for foundation footings must be dug is usually found in the ___ of the foundation plan.
         A. section views
         B. finish schedule
         C. floor plan
         D. plot plan

T   F   **38.** Batter boards are usually placed 4′ to 6′ behind corners to allow working room for form construction.

T   F   **39.** Foundation sills are usually constructed of 2 × 10s or 2 × 12s.

T   F   **40.** Washers should always be used under nuts on anchor bolts.

_____ **41.** A(n) ___ foundation is normally used on a steeply sloped lot.

_____ **42.** The largest part of a concrete mixture consists of fine and coarse ___.

T   F   **43.** Generation is the hardening process that occurs between water and cement in a concrete mixture.

_____ **44.** How many yards of concrete selling for $41.25/yd can be purchased for $660.00?

_____ **45.** A concrete mix contains 1¾ cu ft of gravel, ¾ cu ft of water, 1 cu ft of cement, and 2½ cu ft of sand. What percentage of the mix is gravel?

T   F   **46.** The distance between walers in a wall form is determined by the thickness and height of the wall to be poured.

T   F   **47.** Duplex nails should be used on forms whenever practical.

_____ **48.** Rebars for T-foundation walls are placed ___.
         A. after the outside form walls are placed
         B. before the outside form walls are set
         C. after the inside form walls are set
         D. none of the above

_____ **49.** Sections of a patented panel system are secured to each other with ___.
         A. bolts and nuts
         B. duplex nails
         C. wood screws
         D. wedge bolts or clamps

_____ **50.** After concrete has been placed and struck off, ___ perform finishing operations on concrete slabs.

T   F   **51.** Patios should have a minimum slope of 1″ in 12′ to provide drainage.

_____ **52.** The standard minimum height in a crawl space is ___″ from the bottom of the floor joists to the ground.
   A. 12
   B. 18
   C. 24
   D. 30

_____ **53.** The concrete footings and walls of a low T-foundation are poured ___.
   A. footings first
   B. walls first
   C. at the same time
   D. in no particular order

_____ **54.** Concrete hardens by a chemical action known as ___.

_____ **55.** ___ concrete trucks have the capability of mixing concrete while en route to the job site.

T   F   **56.** Asphalt sheet membranes are sheets of rubberized asphalt laminated to the foundation wall.

_____ **57.** ___ termites are responsible for 95% of all termite damage in the United States.

_____ **58.** Forms for walks and driveways should not be set up until ___.
   A. final soil grading has been completed
   B. shortly before walls are raised
   C. near the end of construction work
   D. none of the above

_____ **59.** The distances from the property lines to the building are known as ___.
   A. setbacks
   B. recesses
   C. offsets
   D. spacings

T   F   **60.** A slab-at-grade foundation features a framed floor unit.

_____ **61.** Building lots must be ___ to determine precise boundaries of a building site.

_____ **62.** ___ foundations are only used for light wall loads and only in firm soil.
   A. Rectangular
   B. T-shaped
   C. Battered
   D. L-shaped

T   F   **63.** Panel forms are generally considered more efficient than built-in-place forms.

T   F   **64.** Foundation footings must be placed below the frost line.

_____ **65.** A(n) ___ line on a plot plan shows the shape of the varying grades of the lot.

Name _____ Date _____

T   F   **1.** When installing floor joists, crowns in the joists should be turned up.

T   F   **2.** Floor trusses for one-family dwellings are normally spaced 16″ OC.

_____ **3.** Wall construction of a house begins after the ___ has been laid.

_____ **4.** A(n) ___ pole may be used in the vertical layout of a wood-framed wall.

T   F   **5.** A nominal 2″ × 4″ stud is 3½″ wide.

_____ **6.** The standard height of interior and exterior doors used in residences is ___.

_____ **7.** Wall studs in residential construction are normally placed ___″ OC.

T   F   **8.** Plywood sheathing may be applied to a squared wall before or after the wall is raised.

T   F   **9.** Fire blocks in exterior walls of one-family dwellings may be installed in a straight line or staggered.

_____ **10.** Ceiling joists should run in the same direction as roof ___ when possible.

T   F   **11.** Lookout rafters on a flat roof are cantilevered.

_____ **12.** The most important framing members of a roof system are the ___.

_____ **13.** Metal structural members may be ___ together or connected with self-drilling flat-head screws.

T   F   **14.** Light-gauge framing receives a protective hot-dip oxidizing treatment.

_____ **15.** Short studded walls in platform construction are known as ___.

_____ **16.** The clear span of a beam is the ___ span.

_____ **17.** A house foundation requires 164′ of 2″ × 4″ sill plate. At $.22½ per lineal foot, what is the cost of the sill plate?

_____ **18.** What is the plywood cost to deck a 24′ × 32′ subfloor with plywood costing $17.47 per sheet?

_____ **19.** ___ joists are located at each side of a floor opening.

T   F   **20.** Load and height are the two major factors affecting the size of joist material needed for a job.

_____ **21.** Edges of standard size subfloor panels should break over the ___ of floor joists.

_____ **22.** Joints between the planks of a built-up beam are ___.

_____ **23.** Header joists are also known as ___ joists.
        A. inner or outer
        B. interior or exterior
        C. rim or band
        D. horizontal or level

T   F   **24.** Diagonal bracing provides lateral strength for a wall.

_____ **25.** Horizontal framing members directly above a stud wall make up the double ___.

_____ **26.** Sheathing materials are attached to a metal-framed wall with ___.

T   F   **27.** The standard height of walls in wood-framed houses is 8'-1½" from subfloor to ceiling joists.

T   F   **28.** Metal fasteners should not be used with wood-framed walls.

T   F   **29.** Stub joists are required in the hip section of a low-pitched hip roof.

_____ **30.** Self-tapping screws include ___ and self-piercing screws.

_____ **31.** ___ is the most commonly used interior wall finish for metal-framed walls.

T   F   **32.** Metal framing members are termite and fire resistant.

T   F   **33.** Beams for a post-and-beam subfloor system are normally placed 48" OC.

_____ **34.** The top and bottom members of a wood floor truss are known as ___.

_____ **35.** A carpenter cuts four 2" × 12" × 14' boards, one 2" × 12" × 10' board, and three 2" × 12" × 12' boards to make a built-up beam of three 2" × 12" pieces 28' long. Has a mistake been made in the cutting?

_____ **36.** Twenty-four sheets of ¾" × 4' × 8' plywood and 112 lineal feet of 2 × 4s are required for the subfloor and sill plate of a house. At $22.32 per sheet for plywood and $.24 per lineal foot for 2 × 4 material, what is the material cost?

T   F   **37.** Floor trusses are generally prefabricated.

_____ **38.** ___ joists run from a header to a supporting wall or beam.

_____ **39.** A(n) ___ is a recess in a foundation wall that receives a beam.

_____ **40.** ___ are nailed into a wall to slow down a fire traveling inside the wall.

_____ **41.** Vertical framing members that run between wall plates are ___.

T   F   **42.** A ⅛" gap should be left between edges of plywood sheathing to allow for possible expansion.

_____ **43.** ___ is the material used for the exterior covering of outside walls.

T   F   **44.** Outside corners can only occur at the end of a wall.

T   F   **45.** When raising wall sections, long sections should be raised last.

T   F   **46.** When roof rafters are spaced 24″ OC and joists are spaced 16″ OC, every other rafter will be placed next to a joist.

T   F   **47.** A flat roof should have a minimum pitch of ¼″ per foot.

T   F   **48.** Both ends of a ceiling joist must rest on an interior wall.

T   F   **49.** After welding galvanized steel components, a corrosion-resistant coating must be applied to the weld area.

T   F   **50.** Floor underlayment is located directly beneath the subfloor.

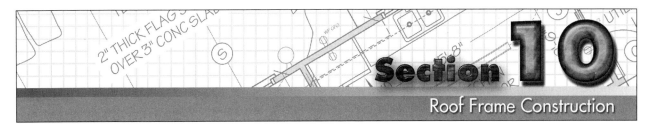
Name _____ Date _____

_____ **1.** What is the total rise of a gable roof with a span of 30′ and a 5″ unit rise?

T   F   **2.** A mansard roof slopes in two directions.

_____ **3.** ___ is the number of inches a rafter rises vertically for each foot of run.

T   F   **4.** Rafter tables are printed on the face side of a steel square.

T   F   **5.** A gable roof slopes in two directions.

T   F   **6.** Ceiling joists for gable roofs are normally spaced 16″ OC.

T   F   **7.** A shed roof slopes in two directions.

T   F   **8.** Rafter tables on a steel square may be used to find the length of hip rafters.

_____ **9.** The side cut of a hip rafter varies with the ___.

_____ **10.** The unit run for a hip rafter is ___″.

_____ **11.** The basic components of a roof truss are ___ and ___.
     A. jacks; studs
     B. chords; web members
     C. ridges; purlins
     D. none of the above

_____ **12.** All parts of a truss are always in a state of ___ or compression.

_____ **13.** ___ and nonveneered panels are often used as roof sheathing.

T   F   **14.** End joints of sheathing panels should be staggered during installation.

_____ **15.** The total ___ of a roof is one-half the total span.

_____ **16.** The slope of a roof is known as the ___.

_____ **17.** *Shortening the rafters* refers to ___.
     A. adding one-half the thickness of the ridge board for each common rafter
     B. adding the thickness of the ridge board for each common rafter
     C. deducting one-half the thickness of the ridge board for each common rafter
     D. deducting the thickness of the ridge board for each common rafter

291

T   F   **18.** A gambrel roof has a double slope on each side.

T   F   **19.** Gable studs provide a nailing surface for siding and sheathing at the end of the roof.

T   F   **20.** Rafters are normally wider than ridge boards.

_____ **21.** Hip rafters run at a(n) ___° angle from the corners of a building to the ridge.

_____ **22.** ___ a hip rafter is faster than backing it.

T   F   **23.** Hip rafters may be dropped by making the seat cut larger.

_____ **24.** Valley rafters are always ___ to hip rafters.
    A. parallel
    B. diagonal
    C. perpendicular
    D. oblique

_____ **25.** Plumb cuts for a common rafter must be laid out on the ___, ___, and ___ of the rafter.

_____ **26.** Fascia boards nailed to the tail ends of ___ rafters serve as a finish piece at the edge of a gable roof.

_____ **27.** The total rise of a roof with a total run of 16′ and a unit rise of 6″ is ___′.

_____ **28.** A(n) ___ roof, which is braced by four rafters, is the strongest roof type.

_____ **29.** Basic roof types can be combined to form ___ roofs.

T   F   **30.** Common rafters for hip roofs are laid out in the same manner as common rafters for gable roofs.

T   F   **31.** Purlins and collar ties are often used to strengthen gable roofs.

_____ **32.** The ___ length of a common rafter is longer than the actual length of the rafter.

T   F   **33.** The overhang for a common rafter is lower than the overhang for a hip rafter.

T   F   **34.** Side cuts are not required on hip rafters.

_____ **35.** A hip jack rafter has plumb cuts ___.
    A. at the heel
    B. at the tail
    C. where it fastens to the hip
    D. all of the above

T   F   **36.** Hip jack rafters increase in length as they near the end of a building.

_____ **37.** A hip rafter must have a(n) ___ cut where it meets the ridge board.

T   F   **38.** Valley cripple jack rafters are used only on intersecting roofs with unequal spans.

_____ **39.** A shortened valley rafter runs at a ___° angle to the supporting valley rafter.
  A. 22½
  B. 45
  C. 60
  D. 90

T F **40.** Over 75% of all new light frame construction uses some type of truss system.

T F **41.** Metal or plywood gusset plates may be used to tie truss parts together.

T F **42.** Camber in a truss helps to offset the tendency of a truss to sag.

T F **43.** A mansard roof has more planes than a gambrel roof.

_____ **44.** The total rise of a roof must be known before the roof ___ can be set at the correct height.

_____ **45.** To find the total rise of a roof, multiply the number of feet in the total run by the ___.

T F **46.** Steeply pitched roofs shed snow more easily than flat roofs.

Name _____ Date _____

T   F   **1.** Heated air is lighter than nonheated air.

T   F   **2.** Wood has a higher k factor than polyurethane foam.

_____ **3.** Heat loss in a residence occurs primarily through the ___.
   A. roof
   B. foundation
   C. walls
   D. doors and windows

_____ **4.** Faced insulation often has a(n) ___ foil vapor barrier.

T   F   **5.** Unfaced blanket and batt insulation stays in place by friction.

T   F   **6.** Foam insulation may be blown into cavities of brick or block walls.

_____ **7.** Sound intensity is expressed in ___.

T   F   **8.** Acoustical tile on a ceiling acts as an effective sound absorber.

T   F   **9.** All solar heating methods require sunlight for efficient operation.

T   F   **10.** Collectors for solar heating are normally placed on the roof of a building.

_____ **11.** Buildings in cold climates require insulation with higher ___ values than insulation used for buildings in warmer climates.

_____ **12.** An open area at the bottom of a thermal storage unit is a(n) ___.

T   F   **13.** Air expands as it is heated.

T   F   **14.** Wood has a higher k factor than concrete.

_____ **15.** ___ is the movement of heat through a solid substance.

_____ **16.** A(n) ___ is the amount of heat required to raise the temperature of 1 lb of water 1°F.

T   F   **17.** Insulators do not transfer heat quickly.

_____ **18.** Insulation is normally placed in new buildings by ___.
   A. carpenters
   B. electricians
   C. plumbers
   D. sheet metal workers

_____ **19.** Materials with a perm rating of ___ or less qualify as vapor barriers.
        A. 0.25
        B. 1.00
        C. 2.50
        D. 5.00

T   F   **20.** Hollow concrete masonry units are a homogenous material.

T   F   **21.** Moisture rises from the ground into a building by radiant induction.

T   F   **22.** Caulking is the best method of sealing small cracks to prevent heat loss.

T   F   **23.** Glass conducts six to ten times more Btu/hr than an equivalent area of framed wall.

T   F   **24.** Roofs should not be insulated with nonstructural rigid insulation panels.

_____ **25.** ___ insulation has high fire-resistive ratings.

T   F   **26.** Insulated glass in windows helps reduce sound transmission.

T   F   **27.** Water-filled containers are used to store heat in the Thrombe wall system.

T   F   **28.** The front wall of a solar-heated house must face south.

T   F   **29.** Active solar heating utilizes mechanical means.

T   F   **30.** Convection is the movement of heat through circulation of air or a liquid.

T   F   **31.** The temperature at which condensation occurs is the dew point.

T   F   **32.** The direct transmission of heat by invisible rays is known as convection.

_____ **33.** An underfloor radiant heating system heats a building by ___.

_____ **34.** A forced, hot air furnace moves heat by ___.

T   F   **35.** Moisture problems due to condensation occur most often in warm climates.

_____ **36.** ___ insulation is the material most widely used to insulate walls, floors, ceilings, and attics.
        A. Blanket and batt
        B. Rigid foam
        C. Loose fill
        D. Foamed-in-place

T   F   **37.** Rigid foam insulation has a higher R value per inch of thickness than any other type of insulation.

T   F   **38.** Foamed-in-place insulation may be poured or sprayed into wall cavities.

T   F   **39.** Heat cannot be transmitted through doors or windows.

T   F   **40.** Loose fill insulation should not be used in wall cavities.

T　　F　　**41.** Acoustical plaster cannot be sprayed on ceilings.

T　　F　　**42.** Breaking the sound vibration of a wall helps to reduce sound transmission through the wall.

T　　F　　**43.** Good insulators are poor conductors.

T　　F　　**44.** Cold air moves toward hot air.

_____ **45.** ___ are any material that conducts heat easily.

T　　F　　**46.** The C factor is used to measure conductivity through non-homogenous materials.

T　　F　　**47.** Heat from a fireplace warms a room by radiation.

_____ **48.** ___ percent of the energy used in an average home is used for heating and cooling.

T　　F　　**49.** Active solar heating systems can be installed only in new buildings.

_____ **50.** To improve sound control, ___ on opposite sides of a hallway should be staggered.

Name _____ Date _____

_____ **1.** The area in which one shingle overlaps a shingle in the course below it is ___.
    A. headlap
    B. toplap
    C. bottomlap
    D. none of the above

T   F   **2.** Exposure refers to the area of a shingle that is not overlapped.

T   F   **3.** The cornice is located directly above the roof overhangs.

T   F   **4.** The top piece of exterior molding on a window is the exterior sill.

_____ **5.** Standard height of interior and exterior doors for a single-family dwelling is ___.

T   F   **6.** Ventilation through a double-hung window cannot exceed 25% of the window opening.

T   F   **7.** Slats on a jalousie window may be opened to 90°.

_____ **8.** ___ is plywood or nonveneered panels under roof shingles.

T   F   **9.** Asphalt shingles have a life expectancy of 20 to 30 years.

T   F   **10.** Frieze boards can be notched between rafters.

T   F   **11.** Soffits are nailed to the sides of rafters to give a finish appearance.

_____ **12.** Roof overhangs are also known as ___.

_____ **13.** Shakes are longer than wood ___.

T   F   **14.** Fixed-sash windows allow more ventilation in a room than double-hung windows.

T   F   **15.** Information pertaining to door and window units of a building is found on the floor plans, elevation plans, and details of a set of prints.

_____ **16.** Sliding glass doors are also known as ___ doors.

T   F   **17.** Casement windows are hinged on two sides.

T   F   **18.** The stool of a window unit is located on the exterior of the window unit.

_____ **19.** A(n) ___ is a roof covering split from a log.

_____ **20.** Spaces between rafters may be left open to provide ventilation to the ___.

_____ **21.** Flashing material used for exterior walls is usually of a(n) ___ material.
    A. galvanized or vinyl
    B. asphalt or cementitious
    C. bituminous or cementitious
    D. none of the above

_____ **22.** ___ sliding windows operate on tracks located at the top and bottom of the window unit.

T   F   **23.** Bevel siding has a tongue on one edge and a groove on the other edge that are used to connect adjoining pieces.

_____ **24.** The area in which one shingle overlaps an adjacent shingle in the same course is known as ___.

_____ **25.** Spacing between adjacent shakes should be ___ to allow for expansion.

T   F   **26.** Built-up roof coverings are normally applied by carpenters.

T   F   **27.** Shakes have a smoother surface than wood shingles.

T   F   **28.** Straightsplit shakes are tapered ¼″ per foot.

T   F   **29.** Shakes are longer than wood shingles.

T   F   **30.** Wood is the oldest type of material used for window units.

T   F   **31.** Window and door units are set in place by carpenters.

_____ **32.** A ___ is not one of the three basic types of overhead garage doors.
    A. sectional roll-up door
    B. one-piece swing-up door
    C. sliding pocket door
    D. rolling steel door

T   F   **33.** Dolly Varden siding must overlap the course below.

_____ **34.** Wood board siding should be ___ with a sealer before being nailed.

_____ **35.** Plywood panels may be applied ___.
    A. directly to studs placed 16″ or 24″ OC
    B. over nonstructural sheathing
    C. either horizontally or vertically
    D. all of the above

_____ **36.** Each pane of glass in a window with muntins is known as a ___.
    A. unipane
    B. sheet
    C. light
    D. none of the above

T   F   **37.** Hardboard may be stained or painted for appearance.

T   F   **38.** Awning windows may be combined with fixed-sash windows in the same opening.

_____**39.** Panel siding can be installed more quickly than ___ siding.

T   F   **40.** Aluminum siding cannot be installed by carpenters.

T   F   **41.** Joints between panels of aluminum siding should be staggered.

T   F   **42.** Rafters must never be cut when installing skylights.

T   F   **43.** Board siding should be applied in a vertical position.

_____**44.** $.375'' + .5'' =$ ___ $''$.

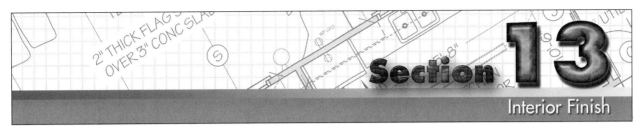

Name _____ Date _____

   T   F   **1.** Rough ceiling height in residential construction is usually 8′-1″.

   T   F   **2.** When hanging drywall, nails should be driven flush with the surface.

_____ **3.** A right-hand reverse door hinges on the ___.
     A. right and opens outward
     B. left and opens outward
     C. right and opens inward
     D. left and opens inward

_____ **4.** Plastic ___ is often used to cover countertops.

   T   F   **5.** A miter joint is preferred over a coped joint for an inside corner of ceiling molding.

   T   F   **6.** Hardwood flooring is less durable than softwood flooring.

   T   F   **7.** Casing is the trim placed around a doorjamb.

   T   F   **8.** Gypsum board should not be applied over metal stud walls.

   T   F   **9.** Hollow core doors provide better sound insulation than solid-core doors.

_____ **10.** ___ corner beads are used to reinforce outside corners of sheetrock walls.

_____ **11.** Scarf joints for base molding are made by overlapping two ___° angles.

_____ **12.** The ½″ space where casing is held short on a door jamb is the ___.

_____ **13.** Sink cut-outs for countertops are usually made ___.
     A. after the cabinet trim is applied
     B. on the job
     C. by the plumber
     D. none of the above

   T   F   **14.** Hollow-core doors are not recommended for outside use.

_____ **15.** ___ is used to cover taped joints and nail dimples.

_____ **16.** An island base cabinet requires a 36″ × 72″ piece of plastic laminate for the countertop. At $3.19/sq ft, what is the material cost of the plastic laminate?

_____ **17.** A straight-run countertop is 122″ long. With a 1″ overhang on each end, ___ and ___ factory-made base cabinets are required.
  A. two 48″; one 36″
  B. one 72″; one 48″
  C. one 72″; one 36″
  D. none of the above

_____ **18.** ___ hinges are used to hang doors in residential and commercial buildings.

_____ **19.** Stile-and-rail doors are also known as ___ doors.

_____ **20.** Exterior doors for residential construction are usually ___″ thick.

_____ **21.** ___ flooring is made of wood strips arranged in various patterns.

_____ **22.** Strip flooring is commercially available in widths ranging from ___″ to ___″.
  A. 1; 3
  B. 1½; 3
  C. 1½; 3½
  D. 2; 4

_____ **23.** A floor that will yield under pressure is said to be ___.

_____ **24.** Nailing blocks must be placed between wall studs when boards are fastened ___ to a wall.

T F **25.** Drywall may be attached to wood studs with screws or nails.

T F **26.** Carpenters may not hang gypsum board.

T F **27.** Base molding is usually applied at the bottom of walls.

_____ **28.** Nails used to install drywall should penetrate at least ___″ into the wood.

_____ **29.** Nine windows in a one-family dwelling require 14′ of casing per window. At $.67/ft, what is the cost of the casing?

_____ **30.** Ceiling tiles used in residential and commercial construction are ___.
  A. fabricated from fiberboard, mineral glass, and similar materials
  B. available in various colors and designs
  C. available with acoustical qualities
  D. all of the above

_____ **31.** A rectangular living room measures 14′-10″ × 20′-6″. ___, ___, and ___ pieces of corner molding should complete the room with the least waste. (Disregard door openings.)
  A. four 8′; two 10′; two 12′
  B. two 14′; four 10′
  C. four 10′; two 4′; two 8′
  D. eight 14′; two 10′; two 8′

T   F   **32.** Joints in drywall should be staggered when possible.

_____**33.** Wood screws should penetrate into the wall ___ when hanging wall cabinets.

_____**34.** The ___ is the rear vertical portion of a countertop.

T   F   **35.** Board-on-board and board-and-batten panels should be applied vertically.

_____**36.** Molding is usually fastened with ___ nails.

_____**37.** When installing single layers of drywall, the long edge is usually applied ___.
    A. vertically
    B. horizontally
    C. with the studs
    D. all of the above

_____**38.** ___ at right angles to each other must be established before installing ceiling tile.

T   F   **39.** Gypsum board has poor fire-resistant qualities.

_____**40.** Wood flooring is commercially available in strips, planks, or ___.

_____**41.** Base molding is installed after the ___ has been nailed in place.

T   F   **42.** Ceiling tiles are normally fastened to furring strips with screws.

_____**43.** A ceiling in a recreation room measures 12'-0" × 16'-0". How many pieces of 2' × 4' tile are required for the ceiling?

_____**44.** A 6'-0" × 10'-0" entryway and a 3'-0" × 8'-0" hallway floor are to be covered with 6" × 6" ceramic tile. At $.98 per tile, what is the tile cost for the job?

T   F   **45.** Subfloors should be covered with waterproof felt or building paper before strip flooring is applied.

_____**46.** When laying floor tiles, the row at the ___ of the room should be laid first.
    A. longest wall
    B. shortest wall
    C. center
    D. none of the above

_____**47.** How many pieces of 1' square floor tile are required to cover the floor of a room measuring 10'-0" × 12'-6"?
    A. 88
    B. 100
    C. 112
    D. none of the above

_____**48.** Panel doors may have plain or ___ panels.

_____**49.** How many hinges should a door over 6'-8" in height have?

T   F   **50.** Vinyl and slate are examples of resilient tile.

_____ **51.** A ___ edge is not a standard edge on a piece of drywall.
        A. tapered
        B. beveled
        C. finger joint
        D. tongue-and-groove

_____ **52.** Panel doors consist of rails, ___, and panels.

T   F   **53.** Magnetic catches cannot be used with flush doors on base cabinets.

T   F   **54.** Contemporary window trim is grooved on the face side for appearance.

T   F   **55.** Gypsum board may be hung with its long dimension running vertically or horizontally.

T   F   **56.** A left-hand reverse door has hinges on the left and opens inward.

_____ **57.** ___ doors are fitted in cabinet openings.

_____ **58.** When installing strip flooring over concrete, nailing strips known as ___ may be used.

T   F   **59.** When nailing strip flooring, nails should be driven at a 45° to 50° angle.

_____ **60.** ___ bolts provide increased security for entrance doors.

T   F   **61.** Nails through strip flooring should penetrate into the floor joists if possible.

T   F   **62.** Hardwood panels should be separated and stacked on their edges at least 48 hr before they are applied to a wall.

_____ **63.** A ceiling is to be finished with 12″ × 12″ acoustical ceiling tiles in a room measuring 14′-0″ × 18′-0″. How many ceiling tiles are required?

T   F   **64.** Battens are required on every other board in board-on-board wall paneling.

Name _____ Date _____

_____ **1.** The most important concern in stairway design is ___.

_____ **2.** Stairways in residential construction are usually ___″ to ___″ wide.
    A. 24; 30
    B. 30; 36
    C. 36; 42
    D. 42; 48

T   F   **3.** An L-shaped stairway runs along three walls.

_____ **4.** Landings must be used to break any stairway rising ___′ or more.
    A. 6
    B. 8
    C. 10
    D. 12

_____ **5.** A(n) ___ stringer is required when one side of the stairway is open.

T   F   **6.** The minimum headway recommended for main stairs in a one-family dwelling is 7′-0″.

T   F   **7.** Handrails on stairs should be at least 36″ high.

_____ **8.** The horizontal surface of a step is the ___.

_____ **9.** Riser heights for exterior stairs should be ___″ to ___″.

T   F   **10.** Stairways in public buildings must be over 52″ wide.

_____ **11.** The vertical distance from one floor to the floor above is the ___.
    A. tread width
    B. stringer height
    C. unit rise
    D. total rise

_____ **12.** A ___ square is used to mark treads and risers on a stringer.
    A. bevel
    B. try
    C. framing
    D. combination

T   F   **13.** U-shaped stairways turn one corner between floors.

_____**14.** A stairway has a tread width of 11″. The preferred riser height is ___″ to ___″.
    A. 4; 5
    B. 5; 6
    C. 6; 7
    D. 7; 8

_____**15.** A(n) ___ is nailed to a stairway header to provide additional support for the stringer.

T   F   **16.** Critical angles are the most comfortable walking angles for a stairway.

_____**17.** The gooseneck of a stair unit is part of the ___.

_____**18.** The tread of a stairway is always placed in a(n) ___ position.

T   F   **19.** Shop-constructed stairways usually have housed stringers.

_____**20.** To find riser height for a stairway, ___.
    A. subtract total rise from total run
    B. divide total rise by number of risers
    C. add total run and number of risers
    D. subtract number of risers from total rise

T   F   **21.** Stringers carry the main load of a staircase.

T   F   **22.** Finish treads and risers are nailed to the stringers.

_____**23.** The distance from one floor to the floor above is 8′-9″. How many risers are required?

_____**24.** Exact tread and riser sizes for stairways are based on the total ___ and run.

_____**25.** A stairway has 7¼″ risers. The tread should not exceed ___″.

_____**26.** The tread projection beyond the face of the riser is known as the ___.

_____**27.** The ___ is the vertical face of a step.

_____**28.** Minimum components required to form stairways are ___.
    A. risers and treads
    B. risers and stringers
    C. treads and stringers
    D. risers, treads, and stringers

T   F   **29.** Straight-flight stairways should not be used for basement stairs.

_____**30.** Riser height of stairways is also known as ___.
    A. total rise
    B. total height
    C. total run
    D. none of the above

T F   **31.** There is always one tread less than the total number of risers in a stairway.

T F   **32.** Thickness of tread material is not considered when determining distance between stringers.

_____ **33.** The horizontal length from the foot of a stairway to the point where the stairway ends above is the ___.
  A. total rise
  B. total run
  C. riser height
  D. tread width

_____ **34.** Treads in stairways are supported by ___.

T F   **35.** Story poles are not used when determining riser heights of stairways.

_____ **36.** ___ provide support when climbing stairs.

_____ **37.** The tread ___ of stairways with winders varies.

T F   **38.** The minimum headroom recommended for service stairs is 6'-8".

_____ **39.** Exterior stairs with closed treads and risers should have a minimum slope of ___" per foot on the treads.

T F   **40.** The balustrade of a stairway is formed by the gooseneck and bullnosing.

_____ **41.** ___ stairways are located on the outside of a building.

T F   **42.** A straight-flight stairway with no landing is the simplest stairway to construct.

T F   **43.** Open stringers are cut to receive treads and risers.

T F   **44.** A starting newel post and landing newel post do not normally occupy the same stairway.

_____ **45.** The minimum recommended space between a handrail and wall is ___".

T F   **46.** Stairways with winders are usually L-shaped.

_____ **47.** Wedges and glue are used to secure treads and risers of ___ stairways.

_____ **48.** A riser is 5¾" high. Will a tread 13¼" wide fall within acceptable tread and riser combinations?

_____ **49.** Wedge-shaped steps in an L-shaped stairway are known as ___.

T F   **50.** Cut-out stringers are used only for interior stairways.

_____ **51.** The width of the tread plus the height of the riser should not equal less than 17" or more than ___".

_____ **52.** The line of travel for a stairway with winders may vary from ___" to ___".

_____ **53.** Stairways in public buildings should be ___" wide or more.

_____**54.** Handrails for stairways must have a cross-sectional width of at least ___″.
    A. 1¼
    B. 1½
    C. 1¾
    D. 2

T   F   **55.** The gooseneck of a stairway is part of the stringer.

_____**56.** ___ for stairs may be cleated, dadoed, cut-out, or housed.

T   F   **57.** The length of a landing should be no less than the width of the stairway.

_____**58.** Center handrails are placed in public stairways over ___″ wide.

T   F   **59.** The open side(s) of any stairway must have handrails.

T   F   **60.** Minimum headroom may be decreased by up to 6″ for parallel flights of stairways.

Name _____ Date _____

T   F   **1.** Both ends of a transverse roof beam rest on a post.

_____   **2.** Sixteen 4″ × 4″ × 10′ posts are required for a small building. At $2150.00 per thousand feet, what is the cost of the posts?

T   F   **3.** Heavy timber structures have more seismic and wind resistance than most other wood-frame structures.

_____   **4.** The normal spacing of wall posts in a post-and-beam building is ___, ___, or ___.

        A. 2′; 4′; 6′
        B. 4′; 6′; 8′
        C. 6′; 8′; 10′
        D. 8′; 10′; 12′

T   F   **5.** Metal fasteners should not be used in post-and-beam construction because they cannot be placed behind finishing materials.

T   F   **6.** Steel connectors are the primary means for tying together the structural components of heavy timber construction.

T   F   **7.** Individual boards of laminated timbers may be glued face-to-face or edge-to-edge.

_____   **8.** A hardwood floor for a post-and-beam building is 12′-0″ × 16′-6″. What is the total number of square feet in the floor?

T   F   **9.** Only tongue-and-groove planks should be used for covering heavy timber roof frames.

T   F   **10.** Timber roof trusses are similar to trusses used in residential construction.

_____   **11.** Plywood floor panels used in post-and-beam construction are usually ___″ thick.

        A. ¾
        B. ⅞
        C. 1
        D. 1⅛

T   F   **12.** Longitudinal roof beams run the full width of a building.

T   F   **13.** The strength and surface appearance of a laminated timber determines its grade.

T   F   **14.** Solid lumber is stronger than glulam timbers of equal size.

T   F   **15.** The maximum load stresses on a glulam beam occur at the top and bottom of a beam.

T   F   **16.** Fiber-reinforced glulam timbers cannot be used for all heavy timber operations.

T   F   **17.** Pressure-treated lumber should be used in pole construction to resist decay.

T   F   **18.** Balanced glulam beams should not be used for cantilevered or continuous span applications.

T   F   **19.** Tongue-and-groove planks may be used to deck the roof of a post-and-beam building.

T   F   **20.** Tall posts without lateral bracing are subject to buckling.

_____ **21.** The minimum size post used in a post-and-beam wall is __″ × __″.

_____ **22.** Each plank of a post-and-beam floor should span at least ___ opening(s) between floor beams.
- A. one
- B. two
- C. three
- D. none of the above

_____ **23.** Post-and-beam construction is also known as ___-and-beam construction.

_____ **24.** The size of beams used in a post-and-beam floor depends upon the ___ between supports.
- A. height and distance
- B. length and width
- C. weight carried and span
- D. none of the above

_____ **25.** Posts in post-and-beam construction are placed in a(n) ___ position.

_____ **26.** Beams in post-and-beam construction are placed in a(n) ___ position.

T   F   **27.** Heavy timbers and stud-framed walls may be combined in heavy timber construction.

T   F   **28.** Veneered beams should not exceed 14′ in length.

_____ **29.** As timber burns, the ___ formed on the wood surface helps to protect the unburned wood.

_____ **30.** Twenty-two posts at $11.18 each are needed for a post-and-beam building. The delivery charge is 5% of the purchase price. What is the total cost of the posts?

_____ **31.** The ___ lams are placed at the top and bottom of a glulam beam.

T   F   **32.** Heavy timber construction is a relatively new building method.

_____ **33.** ___ is the movement of a structural component resulting from stress produced by a heavy applied load.

_____ **34.** Studs used in combination with post-and-beam walls sell for $2.87 each and 94 studs are required to complete a job. What is the cost of the studs?

T   F   **35.** Finger joints should not be used in laminated beams.

_____ **36.** Steel connectors subjected to water penetration are designed with ___ that allow moisture to drain from the connectors.

_____ **37.** Posts are placed 6′ apart in a building of post-and-beam construction. What is the total number of posts required for a 30′ wall, including the corner posts?

T   F   **38.** Truss systems should not be used for roofs of buildings constructed with heavy timber.

T   F   **39.** Poles of pole-constructed buildings may be set in concrete.

T   F   **40.** Metal beams hold their loads longer in a fire than laminated beams.

T   F   **41.** Grain in adjacent boards of a veneered beam should run parallel.

_____ **42.** The most common glulam purlin width used with preframed panelized systems is ___″.

_____ **43.** ___ between butt ends of plank flooring on a post-and-beam floor should be staggered.

T   F   **44.** Outside walls of post-and-beam buildings may be finished with masonry materials.

_____ **45.** ___ portions of a post-and-beam building are usually erected after the outside walls and roof have been completed.

_____ **46.** The two basic post-and-beam roof designs are ___ and ___.

_____ **47.** ___ are constructed using a grid system of beams and purlins tied together with steel connecting devices and hubs.

T   F   **48.** Poles used in pole construction range from 2″ to 12″ in diameter.

T   F   **49.** Insulation materials may be applied to the top side of a post-and-beam roof.

_____ **50.** The material cost for tongue-and-groove flooring on a 16′ × 22′ room is $1144.00. What is the cost per square foot?

_____ **51.** Beams used in floors of post-and-beam construction are usually spaced ___′ on center.
  A. 2
  B. 4
  C. 6
  D. none of the above

_____ **52.** Transverse roof beams run from the outside walls to a(n) ___ beam.

_____ **53.** Timber roof ___ may be fabricated from solid timbers, parallel strand lumber, or glulam.

_____ **54.** For maximum strength, glulam beams are usually placed with the ___ of the lams facing the applied load.

T   F   **55.** Softwood lumber is usually used in the manufacture of laminated timbers.

_____ **56.** ___ is the slight upward curve in a structural member.

T   F   **57.** Three types of post-and-beam construction are residential post-and-beam, timber frame, and beam-frame.

T   F   **58.** Exposed wood beams in post-and-beam construction provide an attractive interior appearance.

T   F   **59.** Metal fasteners are not used in any form of post-and-beam construction.

T   F   **60.** Structural insulated panels should not be used to fill spaces between perimeter wall posts.

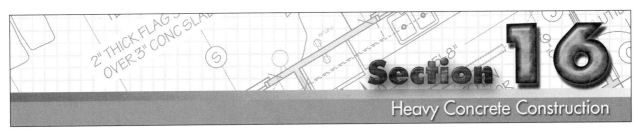
Name _____ Date _____

_____ **1.** A crane has a ¾ cu yd bucket. How many buckets of concrete are required to pour 15 yd of concrete from the bucket?

_____ **2.** One cubic yard of concrete containing 12 lb of water and 20 lb of cement has a water-cement ratio of ___.

_____ **3.** Plyform® panels cost $28.35 each. A contractor uses the panels nine times before discarding them. What is the cost per use?

_____ **4.** A slab is raised 8′ per hour with hydraulic jacks. How long will it take to raise the slab 18′?

T   F   **5.** Piles may be made of wood, steel, or concrete.

_____ **6.** Steel ___ is used to strengthen normally reinforced concrete.

T   F   **7.** Wood is the most widely used building material for concrete forms.

_____ **8.** ___ are often used when it is not possible to excavate the entire site area to the depth required to reach bearing soil.

_____ **9.** ___ holes at the bottom of forms are used to remove debris before the concrete pour.
   A. Weep
   B. Clean-out
   C. Bung
   D. none of the above

_____ **10.** ___ are not a basic structural part of concrete buildings.
   A. Walls
   B. Floors
   C. Roofs
   D. Beams

T   F   **11.** Textured plywood is not used for concrete forms.

_____ **12.** A concrete pour requires 184 cu yd of concrete. The concrete is to be delivered in transit-mix trucks with a capacity of 8 cu yd each. How many truckloads of concrete are required for the pour?

T   F   **13.** Concrete forms should be rigid so that there is no movement during the pour.

315

_____**14.** Precast concrete members may be ___ or bolted together.

T   F   **15.** Leaks in forms cause ridges on the surface of the hardened concrete.

_____**16.** A 1 cu yd bucket will pour the same amount of concrete as a ¾ cu yd bucket in ___% fewer pours.

T   F   **17.** A lift of concrete is also known as a layer.

_____**18.** Mechanical ___ are used to consolidate layers of concrete.

_____**19.** A(n) ___ cone, made of sheet metal, is used to measure the consistency of concrete.

_____**20.** Compression strength is the psi that concrete can withstand ___ days after being placed.

T   F   **21.** A concrete floor slab may bc laid dircctly on the ground.

_____**22.** Eleven pounds of water and 23 lb of cement produce a(n) ___ water-cement ratio.

T   F   **23.** Concrete may expand due to temperature changes.

T   F   **24.** Concrete may expand or contract due to moisture changes.

T   F   **25.** Bullfloats are used to eliminate high and low areas in concrete slabs.

T   F   **26.** Climber tower cranes with buckets are used for pouring operations on large concrete buildings.

T   F   **27.** A structurally supported slab floor is monolithically joined with the walls, columns, and beams.

_____**28.** Curing keeps concrete moist while ___ occurs.

_____**29.** The removal of forms after concrete has set up is known as ___.

_____**30.** The basement and lower floors of a large concrete building are usually poured using a(n) ___ system.

_____**31.** ___ in a concrete mixture binds the sand and gravel together.

_____**32.** Mechanical screeds consolidate and ___ concrete.

T   F   **33.** Carpenters strike off concrete with screeds.

T   F   **34.** Expansion joints in concrete should be removed before concrete has set.

T   F   **35.** Bearing piles are the most commonly used type of piles in concrete heavy construction.

T   F   **36.** Friction piles must penetrate bearing soil to provide proper support.

T   F   **37.** Bored caissons are usually cylindrical in shape.

_____ **38.** ___ foundations, used in soils of low bearing strength, are placed beneath the entire building area.

T   F   **39.** Prefabricated wall panels used in tilt-up construction are usually precast at a plant and delivered to the job site.

_____ **40.** ___-slab construction combines precast concrete or steel columns with floor slabs cast on the job site.

T   F   **41.** Lightweight precast members are raised into place with a crane.

_____ **42.** A layer of concrete is also known as a(n) ___.

_____ **43.** The minor diameter of a slump cone is ___″.

_____ **44.** The major diameter of a slump cone is ___″.

T   F   **45.** Slump cones should be dried thoroughly before use.

T   F   **46.** Control joints may be cut into concrete slabs after the concrete hardens.

T   F   **47.** High-tensile steel cables are placed in prestressed concrete forms.

Appendix

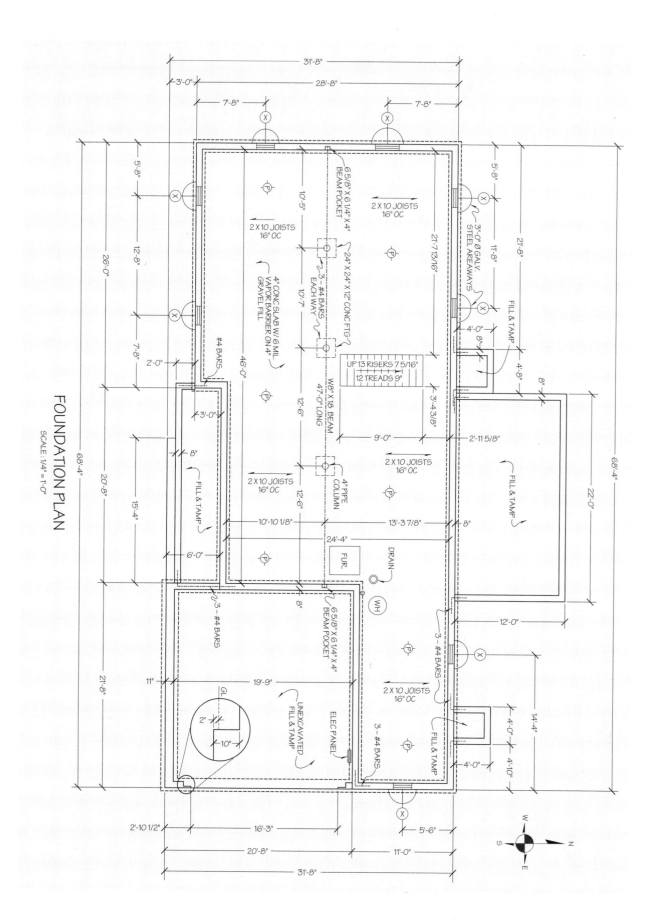

FOUNDATION PLAN

SCALE : 1/4" = 1'-0"

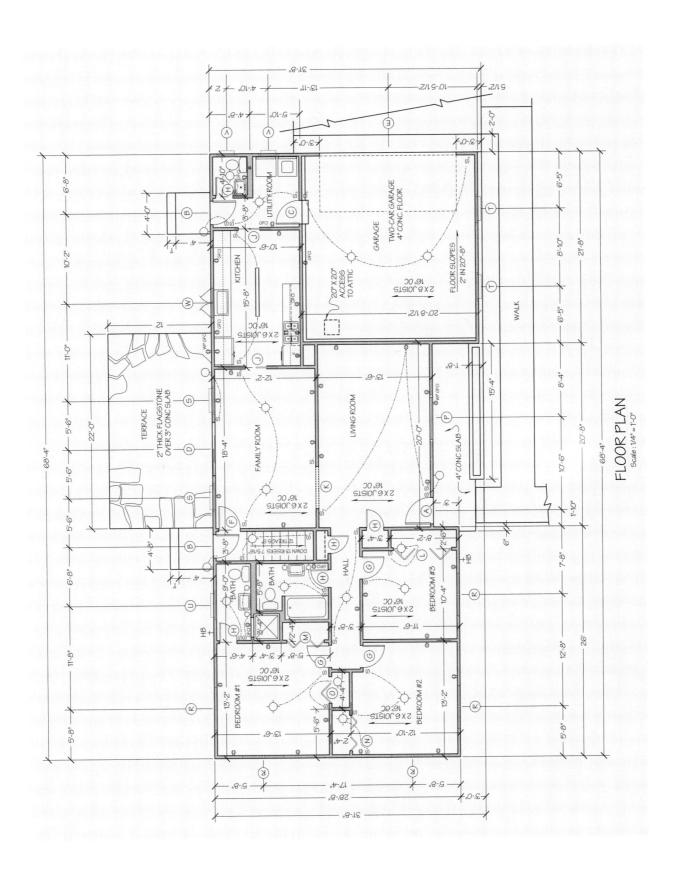

FLOOR PLAN
Scale: 1/4" = 1'-0"

EXTERIOR ELEVATIONS

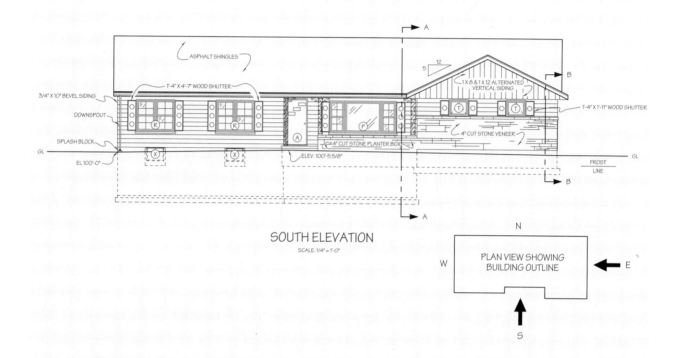

SOUTH ELEVATION
SCALE: 1/4" = 1'-0"

PLAN VIEW SHOWING
BUILDING OUTLINE

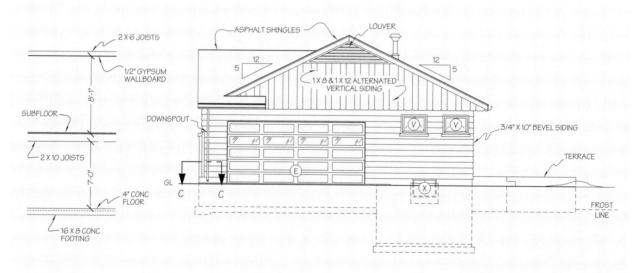

EAST ELEVATION
SCALE: 1/4" = 1'-0"

EXTERIOR ELEVATIONS

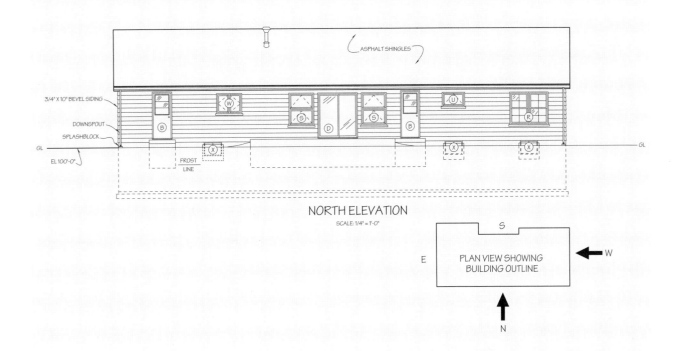

ASPHALT SHINGLES

3/4" X 10" BEVEL SIDING

DOWNSPOUT

SPLASHBLOCK

GL

EL 100'-0"

FROST LINE

NORTH ELEVATION
SCALE: 1/4" = 1'-0"

E

S

PLAN VIEW SHOWING BUILDING OUTLINE

W

N

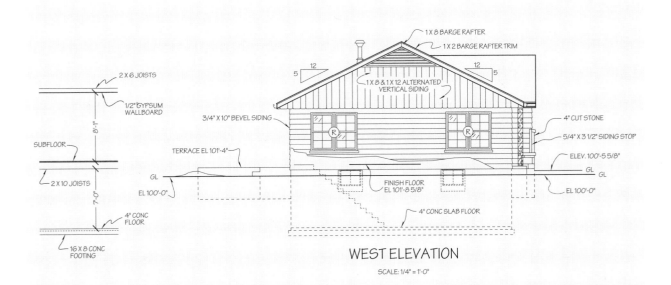

2 X 6 JOISTS

1/2" GYPSUM WALLBOARD

8'-1"

SUBFLOOR

2 X 10 JOISTS

7'-0"

4" CONC FLOOR

16 X 8 CONC FOOTING

1 X 8 BARGE RAFTER

1 X 2 BARGE RAFTER TRIM

12
5

12
5

1 X 8 & 1 X 12 ALTERNATED VERTICAL SIDING

3/4" X 10" BEVEL SIDING

4" CUT STONE

5/4" X 3 1/2" SIDING STOP

TERRACE EL 101'-4"

ELEV. 100'-5 5/8"

GL

GL   GL

EL 100'-0"

FINISH FLOOR EL 101'-8 5/8"

EL 100'-0"

4" CONC SLAB FLOOR

WEST ELEVATION
SCALE: 1/4" = 1'-0"

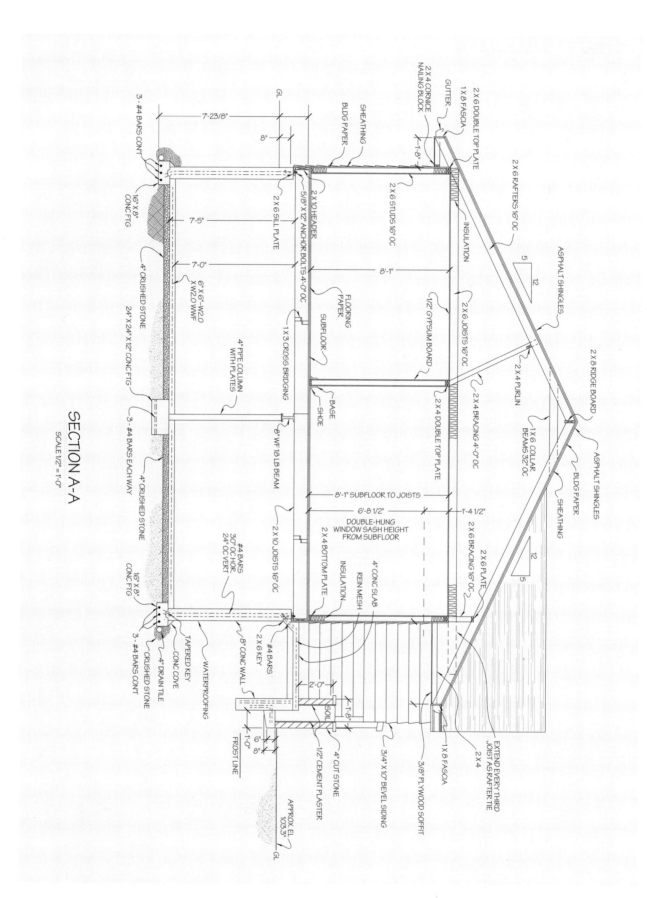

SECTION A-A

SCALE 1/2" = 1'-0"

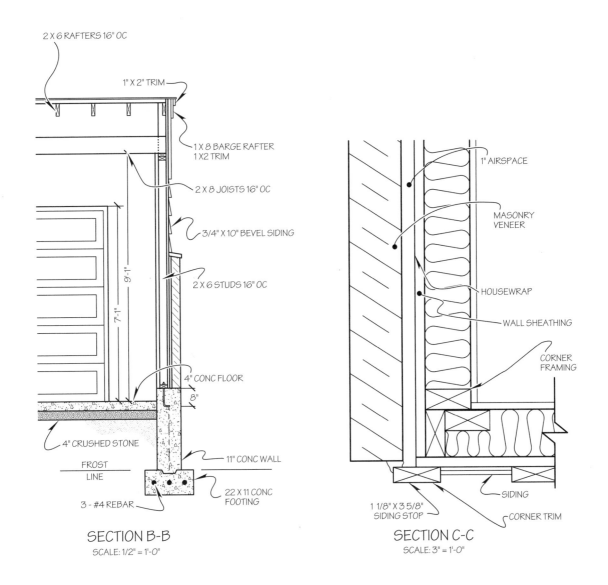

2 X 6 RAFTERS 16" OC

1" X 2" TRIM

1 X 8 BARGE RAFTER
1 X2 TRIM

2 X 8 JOISTS 16" OC

3/4" X 10" BEVEL SIDING

2 X 6 STUDS 16" OC

9'-1"

7'-1"

4" CONC FLOOR

8"

4" CRUSHED STONE

FROST LINE

11" CONC WALL

22 X 11 CONC FOOTING

3 - #4 REBAR

SECTION B-B
SCALE: 1/2" = 1'-0"

1" AIRSPACE

MASONRY VENEER

HOUSEWRAP

WALL SHEATHING

CORNER FRAMING

SIDING

1 1/8" X 3 5/8" SIDING STOP

CORNER TRIM

SECTION C-C
SCALE: 3" = 1'-0"

KITCHEN CABINET ELEVATIONS

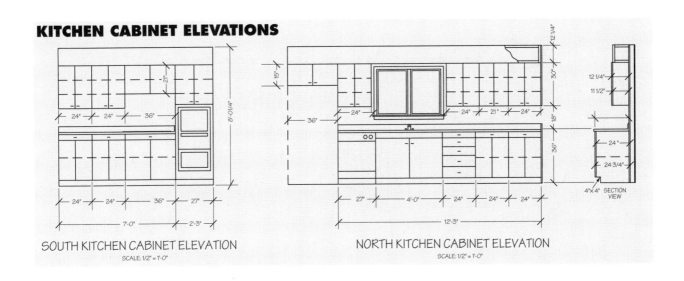

21"

8'-01/4"

24"   24"   36"

24"   24"   36"   27"

7'-0"   2'-3"

SOUTH KITCHEN CABINET ELEVATION
SCALE: 1/2" = 1'-0"

12 1/4"

15"

30"

18"

36"

36"

24"   24"   21"   24"

27"   4'-0"   24"   24"   24"

12'-3"

NORTH KITCHEN CABINET ELEVATION
SCALE: 1/2" = 1'-0"

12 1/4"

11 1/2"

24"

24 3/4"

4" x 4"   SECTION VIEW

FRAMING PLANS

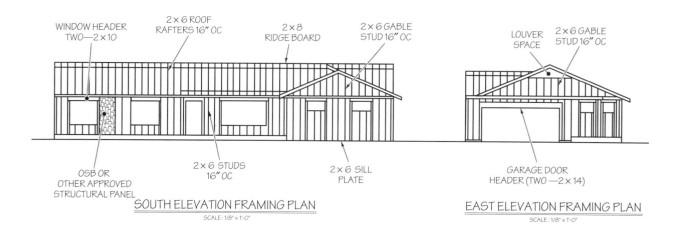

WINDOW HEADER TWO—2 × 10

2 × 6 ROOF RAFTERS 16" OC

2 × 8 RIDGE BOARD

2 × 6 GABLE STUD 16" OC

OSB OR OTHER APPROVED STRUCTURAL PANEL

2 × 6 STUDS 16" OC

2 × 6 SILL PLATE

SOUTH ELEVATION FRAMING PLAN
SCALE : 1/8" = 1'-0"

LOUVER SPACE

2 × 6 GABLE STUD 16" OC

GARAGE DOOR HEADER (TWO —2 × 14)

EAST ELEVATION FRAMING PLAN
SCALE : 1/8" = 1'-0"

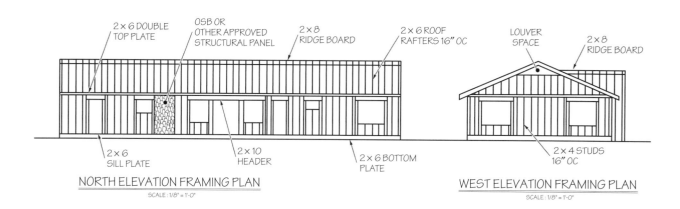

2 × 6 DOUBLE TOP PLATE

OSB OR OTHER APPROVED STRUCTURAL PANEL

2 × 8 RIDGE BOARD

2 × 6 ROOF RAFTERS 16" OC

2 × 6 SILL PLATE

2 × 10 HEADER

2 × 6 BOTTOM PLATE

NORTH ELEVATION FRAMING PLAN
SCALE : 1/8" = 1'-0"

LOUVER SPACE

2 × 8 RIDGE BOARD

2 × 4 STUDS 16" OC

WEST ELEVATION FRAMING PLAN
SCALE : 1/8" = 1'-0"

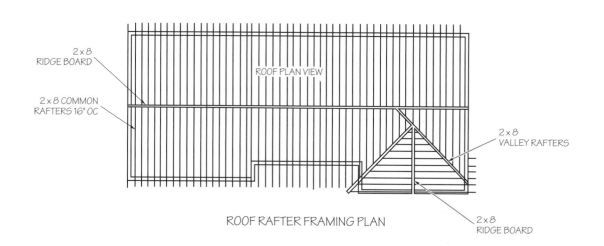

2 × 8 RIDGE BOARD

2 × 8 COMMON RAFTERS 16" OC

ROOF PLAN VIEW

2 × 8 VALLEY RAFTERS

2 × 8 RIDGE BOARD

ROOF RAFTER FRAMING PLAN

FRAMING PLANS

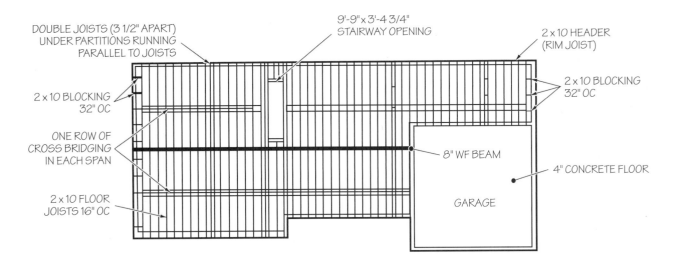

DOUBLE JOISTS (3 1/2" APART) UNDER PARTITIONS RUNNING PARALLEL TO JOISTS

9'-9" x 3'-4 3/4" STAIRWAY OPENING

2 x 10 HEADER (RIM JOIST)

2 x 10 BLOCKING 32" OC

2 x 10 BLOCKING 32" OC

ONE ROW OF CROSS BRIDGING IN EACH SPAN

8" WF BEAM

4" CONCRETE FLOOR

2 x 10 FLOOR JOISTS 16" OC

GARAGE

FLOOR JOIST FRAMING PLAN

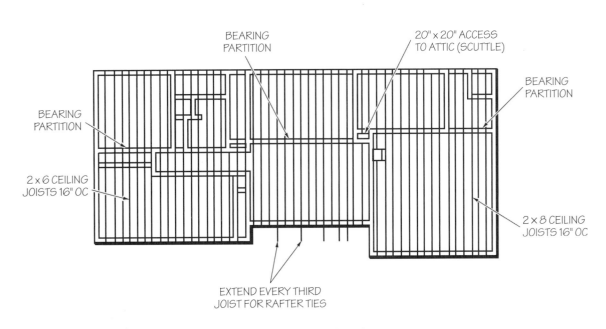

BEARING PARTITION

20" x 20" ACCESS TO ATTIC (SCUTTLE)

BEARING PARTITION

BEARING PARTITION

2 x 6 CEILING JOISTS 16" OC

2 x 8 CEILING JOISTS 16" OC

EXTEND EVERY THIRD JOIST FOR RAFTER TIES

CEILING JOIST FRAMING PLAN

¼″ = 1′-0″ SCALE

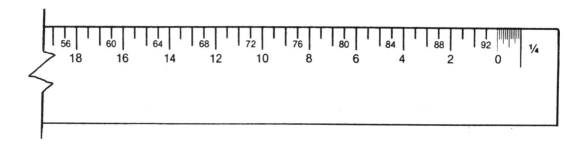

STAIRWAY DETAILS

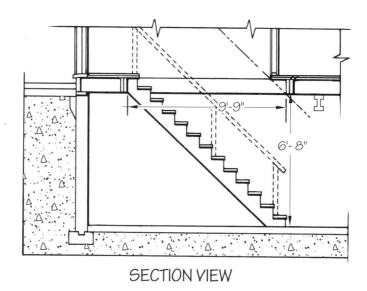

9′-9″

6′-8″

SECTION VIEW

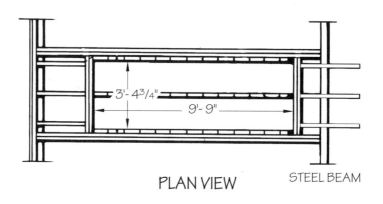

3′-4¾″

9′-9″

PLAN VIEW   STEEL BEAM

DOOR SCHEDULE									
CODE	QUAN	SIZE	THK	ROUGH OPENING	MASONRY OP'G	JAMB SIZE	TYPE	DESIGN	REMARKS
A	1	3′-0″ × 6′-8″	$1\frac{3}{4}$″	3′-2″ × 6′-$10\frac{1}{4}$″		$1\frac{3}{16}$″ × $4\frac{7}{8}$″	HINGED	3 LIGHTS; FLUSH SOLID-CORE	FRONT ENTRANCE DOOR
B	2	2′-8″ × 6′-8″	$1\frac{3}{4}$″	2′-10″ × 6′-$10\frac{1}{4}$″		$1\frac{5}{16}$″ × $4\frac{7}{8}$″	" "	3 LIGHTS; 1 PANEL	REAR ENTRANCE DOORS
C	1	2′-8″ × 6′-8″	$1\frac{3}{8}$″	2′-10″ × 6′-$10\frac{1}{4}$″		$1\frac{5}{16}$″ × $4\frac{5}{8}$″	" "	FLUSH HOLLOW-CORE	DOOR BETWEEN UTILITY ROOM & GARAGE
D	1	6′-$0\frac{1}{8}$″ × 6′-10″		6′-$0\frac{1}{2}$″ × 6′-$10\frac{3}{8}$″			SLIDING	GLASS	SLIDING GLASS DOOR
E	1	16′-0″ × 7′-0″	$1\frac{3}{8}$″	16′-3″ × 7′-$1\frac{1}{2}$″		$\frac{3}{4}$″ × 6″	OVERHEAD	4 LIGHTS; 16 PANELS	GARAGE DOOR
F	1	2′-8″ × 6′-8″	$1\frac{3}{8}$″	2′-10″ × 6′-$10\frac{1}{4}$″		$\frac{3}{4}$″ × $4\frac{5}{8}$″	HINGED	FLUSH HOLLOW-CORE	INTERIOR DOOR
G	3	2′-6″ × 6′-8″	$1\frac{3}{8}$″	2′-8″ × 6′-$10\frac{1}{4}$″		$\frac{3}{4}$″ × $4\frac{5}{8}$″	" "	" "	" "
H	5	2′-0″ × 6′-8″	$1\frac{3}{8}$″	2′-2″ × 6′-$10\frac{1}{4}$″		$\frac{3}{4}$″ × $4\frac{5}{8}$″	" "	" "	" "
J	2	2′-6″ × 6′-8″	$1\frac{3}{8}$″	5′-2″ × 7′-0″			SLIDING	" "	POCKET DOOR
K	1	3′-0″ × 6′-8″		3′-$2\frac{1}{2}$″ × 6′-$10\frac{1}{2}$″		$\frac{3}{4}$″ × $4\frac{5}{8}$″			CASED OPENING
L	1	7′-$7\frac{9}{16}$″ × 7′-11″		7′-$8\frac{9}{16}$″ × 7′-$11\frac{3}{4}$″		$\frac{3}{4}$″ × $4\frac{5}{8}$″	BI-FOLD	FLUSH	BEDROOM #3
M	1	5′-$3\frac{3}{8}$″ × 7′-11″		5′-$4\frac{3}{8}$″ × 7′-$11\frac{3}{4}$″		$\frac{3}{4}$″ × $4\frac{5}{8}$″	" "	" "	BEDROOM #1
N	1	4′-$11\frac{9}{16}$″ × 7′-11″		5′-$0\frac{9}{16}$″ × 7′-$11\frac{3}{4}$″		$\frac{3}{4}$″ × $4\frac{5}{8}$″	" "	" "	BEDROOM #2
O	1	3′-$11\frac{3}{8}$″ × 7′-11″		4′-$0\frac{3}{8}$″ × 7′-$11\frac{3}{4}$″		$\frac{3}{4}$″ × $4\frac{5}{8}$″	" "	" "	BEDROOM #1

WINDOW SCHEDULE							
CODE	QUAN	NO. LTS	GLASS SIZE	SASH SIZE	ROUGH OPENING	MASONRY OP'G	REMARKS
P	1	1 8	56″ × 49″ 20″ × 24″	(1) 5′-0″ × 4′-6″ (2) 2′-0″ × 4′-6″	9′-8″ × 4′-10″		PICTURE WINDOW FLANKED EACH SIDE BY ONE (1) DH WINDOW
R	5	8	28″ × 24″	(2) 2′-8″ × 4′-6″	5′-10″ × 4′-10″		DOUBLE-HUNG WINDOW
S	2	2	39″ × 22″	3′-8″ × 4′-6″	3′-9″ × 4′-$7\frac{3}{16}$″		NO. A-12-84 CURTIS CONVERTIBLE AWNING WINDOW UNIT
T	2	1	39″ × 17″	3′-8″ × 1′-10″	3′-9″ × 1′-$11\frac{3}{8}$″		NO. A-11-83 " " "
U	1	1	33″ × 17″	3′-2″ × 1′-10″	3′-3″ × 1′-$11\frac{3}{8}$″		NO. A-11-73 " " "
V	2	1	27″ × 17″	2′-8″ × 1′-10″	2′-9″ × 1′-$11\frac{3}{8}$″		NO. A-11-63 " " "
W	1	3	17″ × 27″	3′-8″ × 2′-8″	3′-9″ × 2′-$9\frac{3}{8}$″		NO. A-21-56 CURTIS CONVERTIBLE CASEMENT WINDOW UNIT
X	8	2	15″ × 20″	2′-$8\frac{1}{2}$″ × 1′-$10\frac{3}{4}$″		2′-0″ × 1′-11″	STEEL BASEMENT WINDOWS

ABBREVIATIONS

Term	Abbreviation	Term	Abbreviation	Term	Abbreviation
Acoustical Tile	AT or ACT	Escutcheon	ES	Open Web Joist	OJ or OWJ
Air Conditioner	AIR COND	Excavate	EXC	Overhang	OH.
Allowance	ALLOW.	Expanded Metal	EM	Overhead Door	OH. DR
Alternating Current	AC	Exterior	EXT	Panel	PNL
Aluminum	AL	Exterior Grade	EXT GR	Partition	PTN
Anchor Bolt	AB	Face Brick	FB	Penny	d
Application	APPL	Fastener	FSTNR	Pilaster	P
Approximate	APPROX	Fill	F	Plate	PL
Asphalt Roof Shingles	ASPHRS	Finish	FNSH	Plywood	PLYWD
Asphalt Tile	AT.	Finish Floor	FNSH FL	Point of Beginning	POB
Astragal	A	Finish Grade	FG	Porch	P
Auxiliary	AUX	Fireplace	FP	Precast	PRCST
Baseline	BL	Fireproof	FPRF	Pressure-Treated	PT
Basement	BSMT	Fixture	FXTR	Rafter	RFTR
Baseplate	BP	Flashing	FL	Railing	RLG
Bathroom	B	Floor	FL	Redwood	RWD
Bathtub	BT	Flooring	FLG	Refrigerator	REF
Beam	BM	Floor Line	FL	Reinforced	REINF
Bedroom	BR	Footing	FTG	Reinforcing Bar	REBAR
Benchmark	BM	Foundation	FDN	Retaining Wall	RW
Beveled Wood Siding	BWS	Furnace	FURN	Revolving	RVLG
Blanket	BLKT	Furring	FUR.	Ridge	RDG
Block	BLK	Gauge	GA	Right Hand	RH
Board	BD	Galvanized Iron	GI	Riser	R
Board Foot	BF	Garage	GAR	Roll Roofing	RR
Bottom Chord	BC	Girder	GDR	Roof	RF
Brick	BRK	Glass	GL or GLS	Roof Drain	RD
British Thermal Unit	Btu	Grade Line	GL	Roofing	RFG
Building	BL or BLDG	Ground	GND or GRD	Room	RM
Building Line	BL	Ground Fault Circuit Interrupter	GFCI	Rough Opening	RO
Built-up Roofing	BUR			Rough Sawn	RS
Cabinet	CAB.	Gutter	GUT.	Screen	SCR
Cantilever	CANTIL	Gypsum Board	GYP BD	Sewer	SEW.
Casement	CSMT	Hardboard	HBD	Shake	SHK
Cedar	CDR	Hardware	HDW	Sheathing	SHTHG
Ceiling	CLG	Hardwood	HDWD	Sheet Metal	SM
Cement	CEM	Head	HD	Shingle	SHGL
Center	CTR	Header	HDR	Shower	SH
Centerline	CL	Heat	H or HT	Side Light	SI LT
Ceramic Tile	CT	Heating, Ventilating, and Air Conditioning	HVAC	Siding	SDG
Chimney	CHM			Sill	SL
Chord	CHD	Height	HGT	Sink	S or SK
Circle	CIR	Hinge	HNG	Skylight	SLT
Cleat	CLT	Hollow Core	HC	Sliding Door	SL DR
Closet	CLO	Hose Bibb	HB	Soil Pipe	SP
Column	COL	Inch	IN.	Square Foot	SQ FT
Concealed	CNCL	Insulation	INS or INSUL	Square Inch	SQ IN.
Concrete	CONC	Interior	INT	Stack Vent	SV
Concrete Block	CONC BLK	Jamb	JB or JMB	Stainless Steel	SST
Continuous	CONT	Joist	J or JST	Steel	STL
Cornice	COR	Keyway	KWY	Stone	STN
Corrugated	CORR	Kiln-Dried	KD	Strongback	STRBK
Counterbore	CBORE	Kitchen	K or KIT.	Temperature	TEMP
Counterflashing	CFLG	Laundry	LAU	Thick	THK
Countersink	CSK	Lavatory	LAV	Threshold	TH
Critical Path Method	CPM	Left Hand	LH	Tongue-and-Groove	T&G
Cubic Yard	CU YD	Level	LVL	Top Hinged	TH
Dampproofing	DP	Light	LT	Track	TR
Decibel	DB	Linen Closet	LC or LCL	Tread	TR
Detail	DET or DTL	Linoleum	LINO	Truss	TR
Diameter	D or DIA	Lintel	LNTL	Typical	TYP
Dimension	DIM.	Living Room	LR	Utility Room	URM
Dining Room	DR	Lookout	LKT	Valley	VAL
Dishwasher	DW	Louver	LV or LVR	Ventilation	VENT.
Door	DR	Lumber	LBR	Vent Stack	VS
Door Closer	DCL	Masonry	MSNRY	Vinyl Tile	VA TILE
Door Stop	DST	Maximum	MAX.	Washing Machine	WM
Double	DBL	Membrane Waterproofing	MWP	Water Closet	WC
Double Acting	DBL ACT.	Metal	MET. or MTL	Water Heater	WH
Double-Acting Door	DAD.	Metal Flashing	METF	Waterproof	WP
Double Hung Window	DHW	Metal Threshold	MT	Water-Resistant	WR
Douglas Fir	DF	Minimum	MIN	Weatherstripping	WS
Downspout	DS	Miter	MIT	Welded Wire Reinforcement	WWR
Drain	DR	Motor-Operated	MO	West	W
Dressed (lumber)	DRS	North	N	Wide	W
Drywall	DW	Nosing	NOS	Wide Flange	WF
East	E	Not to Scale	NTS	Without	W/O
Electric	ELEC	On Center	OC	Wood	WD
Elevation	EL	Opening	OPNG		